CFZ Yearbook
1996

Jonathan Downes

Typeset by Jonathan Downes, Scanned from the original bt Oll Lewis
Cover and Layout by Hennis and OrangeKat for CFZ Communications
Using Microsoft Word 2000, Microsoft , Publisher 2000, Adobe Photoshop CS.

Photographs © 2007 CFZ except where noted

First published in Great Britain by CFZ Press

**CFZ Press
Myrtle Cottage
Woolsery
Bideford
North Devon
EX39 5QR**

© CFZ MMVIIII

All rights reserved. Without limiting the rights under copyright reserved above, no part of this publication may be reproduced, stored in or introduced into a retrieval system, or transmitted, in any form of by any means (electronic, mechanical, photocopying, recording or otherwise), without the prior written permission of both the copyright owners and the publishers of this book.

ISBN: 978-1-905723-22-5

INTRODUCTION

To the 2008 reprint

Dear friends,

An awful lot of water has flowed under an awful lot of bridges since I last worked on the CFZ Yearbook 1996. The CFZ was about three years old, and we had been publishing *Animals & Men* for about half that time. A new office superstore called *Staples* had just opened up in Exeter, and as part of their ongoing price war against their competitors they were offering bulk photocopying at a ludicrously low rate.

I had one of those bright ideas that have peppered my professional life. They usually come out of nowhere, often are disastrously stupid, and occasionally turn out to be very good ones. This was one of the latter.

At the time *Animals & Men* was only about 32 pages long, and we were receiving far more material than we could ever hope to include in the magazine. I was also very aware that at the time the CFZ was really only my first wife, about a hundred subscribers, and a smattering of hangers-on, and me. We were – as I claimed – the biggest general cryptozoological organisation in the world at the time, but that wasn't actually saying much. The ISC had effectively ceased to exist, and there was nobody else apart from the British Columbia Scientific Cryptozoology Club who were doing a fine job, but were regionally based. So, I wanted to push us up a level by producing a second regular publication .

So the yearbook was born out of a mixture of blind ambition and cheap photocopying. Since then there have been another nine!

Sadly, the original master copies of the yearbook have disappeared during the intervening years. At the time of writing I am still in the midst of moving the remainder of the CFZ archives from Exeter up to North Devon, and no-doubt they are hidden somewhere amongst this wallage of files. But it will take years to sort through them all, so – rather than have this valuable volume out of print for an unspecified length of time – we decided to remaster it from the masters from which the photocopied versions that we have been selling for the past ten years were made. This we have done, and with a few minor bits of tweaking aside, this volume is a facsimile of the first 1995 edition, typos and all.

I hope that you enjoy it.

Re-reading it for the first time in over a decade, I am really impressed with what we had achieved. Without any of the technical advances of the past few years (this book was compiled with an old Amiga 500+ games computer, a dodgy dot-matrix printer, letraset and glue), we still produced a bloody good volume. It includes important research papers by Dr Karl Shuker, Richard Muirhead, and Michel Raynal, amongst others.

However, the most important thing as far as I am concerned is that so many people who were involved in the CFZ back then are still here today. Sadly, my first wife Alison and I divorced less than a year after this volume appeared, Jan Williams left the fold soon after, and I haven't heard from Tom Anderson, Justin Boote, Alistair Curzon, Francois de Sarre and several others for years, but Karl Shuker, Richard Muirhead and Tony Shiels are still involved and are all close personal friends to this day.

The CFZ is now sixteen years old. It has been a bumpy old ride, and will continue to be, but I hope that you agree that it has been worth it.

Slainte

Jon Downes
(Director CFZ)
Woolfardisworthy, North Devon
February 2008

CONTENTS

3. Introduction
5. Contents
6. Introduction to the original edition
7. Who's who in the Centre for Fortean Zoology
8. *Sky Beasts and Cloud Creatures - fiction or fact?* By Dr. Karl P.N.Shuker
26. *The Nnidnidifcication of Ness* by Tony `Doc` Shiels
29. *The riddle of the Sea-Eagle - a cryptic raptor from the westcountry* by Jon Downes
49. *Catmap - a software package to model and study spatial and temporal distribution of Alien Big Cat reports* by Aliastair Curzon
65. *Black Shuck of Norfolk-a poem* by Noella McKenzie
66. *The mystery bird from Hiva-Oa* by Michel Raynal
78. *Big Cats in Norfolk during 1995* by Justin Boote
79. *When the boat comes in - unusual fish landed by the Fleetwood fishing fleet between 1897 and 1965* by Stuart Leadbetter
87. The Beasts of Kingsteignton - an investigation by the Centre for Fortean Zoology
101. *About the survival of relict hominids from a zoologist's point of view* by Francois de Sarre
115. *The Flying Snake of Namibia - an investigation* by Richard Muirhead
127. Book Reviews by Hermann Reichenbach and Jonathan Downes
130. *Loch Ness - a cauldron of definites and possibilities* by Neil Arnold
135. *Notes on the whaling industry at Peterhead* by Tom Anderson
139. *The Marauder of Talog* by Ian Hawthorne together with notes on feral dogs in Britain by Jonathan Downes
142. *Mystery cats in Scotland during 1995* by Tom Anderson
144. *Cryptomicromammology in Europe* by Patrick Brunet-Lacompte
150. THREE ASPECTS OF DURER'S RHINO
    - *A trip to see the rhino* by Noella Mackenzie
    - *Early rhinoceroses in Europe* by Clinton Keeling
    - Notes on the unexpected morphology of Durer's rhino
154. *Do dinosaurs still walk the earth?* By Roy Kerridge
161. A tribute to Jane Bradley
170. *Was an Onza shot in early 1995?* By Dr. Rafael A Lara-Palmeros
172. *The story of a strange fish - quasi fortean aspects to the discovery of the coelacanth* by Jonathan Downes
176. A complete index to *Animals & Men* issues 1-7
    - Index of authors
    - Index of subjects covered
    - Index of articles and letters by title
187. 1995 - a year in the life of the Centre for Fortean Zoology

INTRODUCTION

Welcome to the first edition of a new venture for us at the Centre for Fortean Zoology. 'Animals & Men' is a quarterly magazine which contains short articles and news items on all subjects of interest to the fortean zoologist. We have been receiving articles of greater length, and we did not feel that it was appropriate to split them. We also needed a forum for our own research papers, and whilst we toyed with the idea of making the magazine itself thicker, or more frequent, we felt that to do so would necessarily increase our subscription charges to such a degree that it would not be financially viable to do so.

In July 1995 we distributed a questionnaire to gauge people's response to what we did. One of the questions was "Do you think the contents of the magazine is academic enough?"

If you want to know who to blame for the publication of this yearbook, the answer is simple. Blame Mike Grayson. Mike is one of our readers and he sent us such a nice letter to accompany his questionnaire which said that he 'didn't think we were academic enough', that we decided to do something especially for him. That is not quite true...but its almost true...like so many things in Fortean Zoology!

Mike wrote the above mentioned letter to us in August, and we met him, quite by chance at Zoologica in early September. He told us that he was happy with the magazine but wished that there were more lengthy research papers, like the one about Pine Martens that we had printed in issue one.

We decided, therefore, to print a regular year-book containing research papers and other material that for various reasons was not suitable for inclusion in 'Animals & Men'. We have tried to include material from across the spectrum of fortean zoology. All subjects have been covered, and the contributors range from the extremely well known to the enthusiastic amateur.

Not all the material presented here can be described as 'academic', although a large proportion is so. Some of the material was included just because we liked it. We hope that we have got the balance of academic and popular, scientific and weird just about right, and that whatever else this book provides all its readers with an entertaining miscellany of reading matter. We hope that the material presented here will stimulate discussion, promote further research and provide a bench-mark, so that we, at least, can do better next year!

Jonathan Downes,
Exeter, October 1995

The Centre for Fortean Zoology

15 HOLNE COURT, EXWICK, EXETER EX4 2NA

Jonathan Downes: Editor
Jan Williams: Newsfile Editor
Alison Downes: Administratrix
Lisa Peach: Artist
Mort D Arthur: Cartoons
John Jacques: Sole Representation
Graham Inglis: 'We are the road crew'

CONSULTANTS

Dr Bernard Heuvelmans
(Honorary Consulting Editor)
Dr Karl Shuker
(Cryptozoological Consultant)
C.H.Keeling
(Zoological Consultant)
Tony 'Doc' Shiels
(Surrealchemist in Residence)

Regional Representatives

West Midlands: Dr Karl Shuker.
Mexico: Dr Lara Palmeros
Spain: Angel Morant Fores & Alberto Lopez Acha
Germany: Wolfgang Schmidt and Hermann Reichenbach
France: Francois de Sarre
Wiltshire: Richard Muirhead
Scotland: Tom Anderson
Kent: Neil Arnold
Sussex: Sally Parsons
Hampshire: Darren Naish
Lancashire: Stuart Leadbetter
Belgium: ABEPAR
Norfolk: Justin Boote
North Yorks: Alistair Curzon
Cumbria: Brian Goodwin
Home Counties; Philip Kiberd
S Wales/Salop: John Matthias
Denmark: Lars Thoma and Eric Sorenson
Eire: The wizard of the Western World

Sky Beasts and Cloud Creatures-
Fiction or Fact?

by Dr Karl P.N.Shuker

'None knowes their nest, none knowes the dam that breeds them;
Foodless they live, for th'aire alonely feeds them;
wingless they fly; and yet their flight extends
Till with their flight their unknown lives-date ends'.

Guillaume Sakkuste du Bartas - Divine Weekes and Workes.

'He can only behold,
with unaffrighted eyes,
the horrors of the deep,
and terrors of the skies'.

Thomas Campion - A Book of Airs, xviii

There are few habitats on land or in the water that do not sustain some form of animal life. From mountains and meadows, deserts and jungles, to glaciers, rivers, lakes and from the shallowest seashore rockpools to the cold, lightless abysses and warm hydrothermal vents on the ocean floor -all possess characteristic fauna modified to withstand and utilise the prevailing environmental conditions. But what about the world above the land and water - the sky? Although many species of winged creature - from birds and bats to a multitude of insects, and also many gliding beasts - do indeed spend much of their lives in aerial activity, nature has yet to unfurl animals that spend their entire lives in the sky and clouds continually airborne.

This conspicuous hiatus in the otherwise comprehensive catalogue of habitats and complimentary fauna may explain why so many ostensibly-separate mysterious and controversial phenomena, past and present, have been linked at one time or another with the postulated existence (unknown to us or unexplained by science), of bona fide sky-beasts and cloud creatures - as demonstrated by the following survey's selection of examples.

The Miraculous Manucodiata.

Swifts are amongst the most specialised of birds and some species are known to spend up to nine months of every year in continuous flight - eating, drinking, resting, even mating on the wing - descending from the skies only to build nests and lay eggs. Not without reason therefore, are swifts claimed by some authorities to be the most aerial birds in the world *(Animals May 1970)*. Nevertheless, for more than two-and-a-half centuries Western scientists fervently believed in the existence of extraordinary birds infinitely more aerial than even the most accomplished of swifts - for these were birds that lived their complete lives, from birth to death, drifting delicately through the heavens, and so perfectly adapted for this ethereal existence that they possessed neither bones nor blood, and neither flesh nor feet!

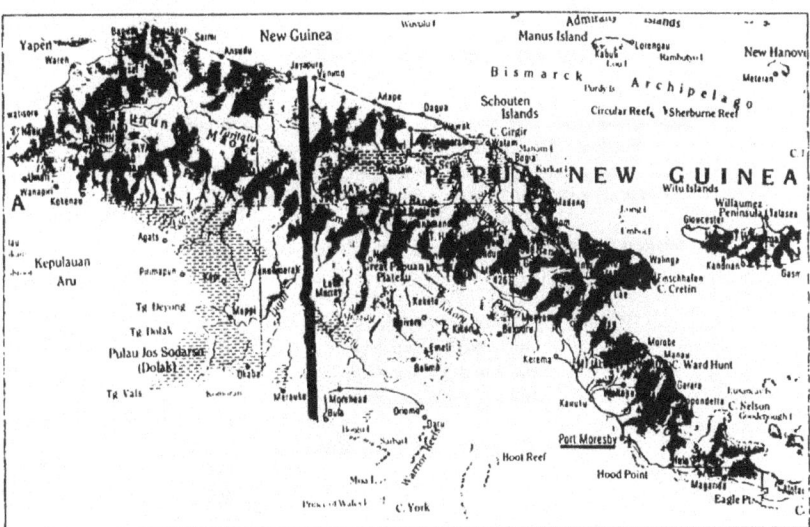

The island of New-Guinea

In September 1522 the surviving vessel of the once-mighty expeditionary fleet of renowned explorer Ferdinand Magellan arrived back home in Seville, bringing with it all manner of exotic treasures and relics from far-flung corners of the globe. Among these was a series of truly exceptional bird skins, which had been purchased from natives of New Guinea and various of its outlying islands. their most immediately striking features were their extravagantly flamboyant feathers - spectacular flourishes of gauzey, rainbow-hued plumes that billowed like dazzling fountains from beneath their wings and tail. When examined more closely, however, these resplendent specimens revealed an even more remarkable characteristic - they were wholly devoid of flesh, blood and bones. Their heads came complete with eyes and a beak, and their bodies had wings, but otherwise it seemed that these extraordinary birds were composed entirely of feathers - they did not even possess any feet! Yet, there were no recognisable signs that the skins had been in any way tampered with, so the possibility of a hoax was discounted.

The belief in fabulous sylph-like creatures such as these recurs over and over in mythology throughout the world, but never before had science obtained any hard evidence in support of their reality. Needless to say, therefore, zoologists were totally bemused, but at the same time thoroughly captivated by these astonishing specimens, and concluded from their near weightless, fleshless, and footless forms, that they undoubtedly lived an exclusively aerial

'Birds of Paradise - once thought to be footless, boneless beings that lived in the skies'.

Picture Courtesy Dr Shuker

existence, presumably sustained solely upon an ambrosial diet of nectar and dew imbibed in flight. To quote one zoologist of that time, they were nothing less than .... *"higher beings, free from the necessity of all other creatures to touch the ground"*. Not surprisingly, as birds that seemed to have originated from Paradise itself, their species ultimately became known as the bird of paradise, and as the manucodiata ('bird of God').

Expeditions to New Guinea brought back more skins, again purchased directly from native tribes, and it soon became obvious that these exquisite creatures comprised many different species, distinguished from one another by their distinct but all equally splendid plumages. No living specimens, however, were captured, and it was not until the 19th Century that Western scientists penetrated the dark New Guinea junglelands to spy these gorgeous birds for themselves - whereupon they finally uncovered their long-secluded secret.

The skins that had been arriving back in Europe were incomplete ones - the New Guinea natives had developed to a fine art the immensely skilled process of skin preparation whereby the flesh, blood, bones and feet of these birds were removed without leaving behind any readily-noticeable signs of their former presence. In short, the birds of paradise were not ethereal, everlastingly-airborne beings at all - in fact, as ornithologists swiftly discovered when able at last to examine complete specimens, they were nothing more than gaudy neighbours of the sombrely-plumaged rooks and ravens. Happily, however, their wonderful feathers were genuine, therefore offering at least a measure of consolation and compensation to scientists and poets alike for the otherwise traumatic transformation of the miraculous manucodiata into first-cousins (albeit very beautiful ones) of the crow family.

Frog Rain and Angel Hair.

One mystery that is still unsolved is the startling phenomenon of animal falls - showers of rain that contain diverse numbers of normally non-aerial animal forms, including worms, toads, fishes and, in particular, frogs. Reported from the very earliest times right up to the present day, a given animal fall generally contains only a single species, but the number of specimens within it is very variable.

Over the centuries, some of the many theories propounded to explain falls of animals from above have aired the)(possibility that the beasts engaged in this perplexing precipitation have actually been born (or have at least spent a considerable time) in the skies. Early philosophers in particular saw little problem with such a concept. believing as they did that frogs, worms, and various other 'creeping things' spontaneously generated from mud, slime, dust, or even dead flesh (the now obsolete theory of abiogenesis), it was simplicity itself to explain their presence in showers of rain as a consequence of spontaneous generation from airborne dust particles, or from raindrops.

In modern times, albeit semi-humorously, Charles Fort envisaged a celestial Sargasso Sea, where creatures lifted into the air by whirlwinds, waterspouts or similar activity remained aloft until metereological disturbances such as storms dislodged them., so they cascaded back down to earth amid the ensuing rain. Taking this scenario even further, others have considered the possibility that perhaps if such creatures were sucked into the air when only very you, they might conceivably mature into adulthood before their eventual descent. With frogs, for example, this hypothesis could predict that frogspawn lifted from pools during a whirlwind could hatch into tadpoles, whilst in the clouds, perhaps even metamorphosing into fully-formed frogs before plummeting back to the ground.

Of course, such an occurrence is a highly improbable one, to say the very least, but it must also be said that even the more reasonable, plausible theories on offer are distinctly less than wholly satisfactory too. Even the whirlwind/waterspout idea, to explain the animals' presence in the air to begin with is fatally flawed. After all, how can it account for the remarkable selectivity of animal falls? Inexplicably, a given fall usually contains just a given species (yet pools or undergrowth usually harbour numerous different types of animal); and the falling animals are rarely accompanied by mud, soil or plant material (yet these would surely be scooped up with them by a whirlwind or waterspout).

Faced with such bewildering paradoxes, it is little wonder, perhaps, that some scientists have sought to eradicate the whole issue of animal falls by saying that they are nothing more than illusions. That is, that they maintain that the animals were never in the sky at all,; instead they were merely lurking in undergrowth or soil (in the case of frogs, worms etc), and were induced by the pleasant moisture of a rain shower to emerge rapidly en masse, creating the illusion that they had fallen from the sky with the rain. And a fish filled pond overflowing during heavy rain is offered as the equivalent 'illusion' for falls of fishes.

Although this theory may well explain certain cases on record, it definitely cannot explain all of them - as demonstrated vehemently by an incident experienced by two members of my own family. In or around 1902, when my maternal Grandmother, Mrs Gertrude Timmins (nee Griffin) was about eight yerars old, she was walking with her mother Mrs Mary Griffin, across some fields in what is now the English town of West Bromwich, in the West Midlands, when it began to rain. Accordingly, they put up their umbrellas and carried on walking, but as they did so, my grandmother was alarmed to feel a series of quite heavy thumps on top of her umbrella - as if pebbles or some other kind of small object were dropping on to it from above. To her surprise, and even greater alarm, when she peeked out from under it she saw that the 'objects' were tiny frogs - tumbling downwards in great numbers, hitting her umbrella, (and also her mother's), and bouncing off the ground, where, apparently unharmed, they hopped away. This fall of frogs continued for several minutes, but her fears were calmed by her mother, who informed her in a totally matter-of-fact, unconcerned manner that frog rains were quite common but completely harmless.

Eventually the rain stopped, and so did the descent of the frogs, after which my grandmother cautiously lowered her brolly, shook off a few frogs still on it, and continued on her way home with her mother. My grandmother died in 1994, at the age of 99, but the incident of the frog rain made such as impression on her that throughout her life she could readily recall all of it. She always stated categorically that the frogs were definitely falling from the sky, rather than merely issuing forth out of the grass around her feet; there were no nearby buildings or trees from which the frogs could have been falling either.

Very worthy of mention here is that in his *Verbreitungsgeschichte der Susswassertierwelt Europas (1950)*, dealing with Europe's freshwater fauna, Dr A. Thienemann has recorded a number of central European cases in which transportation via whirlwind, cyclone, or suchlike provides the only explanation for the confirmed colonisation by frogs and fishes of completely isolated or inaccesible expanses of freshwater.

Also of great interest: by fitting with fine nets various ocean-going ships and also aeroplanes crossing the south Pacific, Dr J.L.Gressitt has revealed that the ostensibly empty air far above the ocean surface actually provides a most unexpected parallel with the latter - by containing a rich aerial 'plankton' of animal life, not only including numerous types of winged insect but also many traditionally sedentary creatures, such as spiders, still attached to their webs and swiftly propelled along by currents of wind in dramatic displays of incongruous hang-gliding. Needless to say, transportation of this type could readily explain the colonisation of isolated, newly-emerged oceanic islets with invertebrate life.

Moreover, some spiders are noted for their tendency to secrete long strands of cobweb and then, while firmly attached to them, to cast their gossamer-like parachutes into the wind in a deliberate attempt to harness its transportation powers, enabling these wingless creatures to be carried through the air across great distances. Indeed, some researchers believe that windborne spider webs and cobweb strands may be the identity of an enigmatic fibrillose substance popularly termed 'angel hair', occasionally reported falling from the sky.

'What is the true explanation for falls of frogs from the skies?'
Picture courtesy of Dr. Shuker.

Nevertheless, much mystery still awaits a solution concerning animal falls from the skies. Even if whirlwinds and similar forces can transport frogs and the like, how can their notable selectivity be explained? Furthermore, once such creatures do become aloft in the skies, how long can they stay there before returning, still alive, to the ground? Many centuries have passed since this bewildering phenomenon was first documented, but we are as far away as ever from obtaining a satisfactory solution to it.

Cloud-Medusas and Sky-Serpents.

The existence of giant aerial creatures on other worlds has been extensively exploited in science-fiction works, engendering a vast diversity of forms encapsulated in numerous novels, short stories, films and television programmes. Perhaps the most famous of all such example   Arthur C.Clarke's novella *'A Meeting with Medusa'* (1971), which features an cr encounter by humans with an immense jellyfish-like beast dwelling among the clouds of Jupiter's atmosphere. 'A Meeting With Medusa' (1971), which features an encounter by humans with an immense jellyfish-like beast dwelling among the clouds of Jupiter's atmosphere. 'A Meeting With Medusa' deservedly won the coveted Nebula Award for Best Novella in 1972, but its impact was far from overeven then, for within a few years its science-fiction scenario would transform into a science-fact hypothesis.

'Do enormous jellyfish-like creatures dwell among the clouds of Jupiter?'
Picture courtesy of Dr. Shuker

In 1977, Cornell University astrophysicist Dr E.E. Salpeter and biologist-astronomer Dr Carl Sagan joined forces to publish a paper in the Astrophysical Journal Supplement that set out their thoughts concerning the possible presence of life on Jupiter, and the likely form(s) that it may take, existing amidst this gargantuan gas planet's clouds. They concluded that much life as might reside there may well parallel the ecology of Earth's ocean fauna, thereby yielding Jovian equivalents to our plankton, our fishes, and our larger, fish-eating sea-life (turtles, giant fishes, etc) Salpeter and Sagan postulated that they would be enormous sac-like organisms, filled with helium, and propelling themselves through the planet's atmosphere by controlled expulsion of this gas from their bodies.

Furthermore, the two scientists did not merely consider these airborne, balloon-like beasts to be wholly conjectural. On the contrary, they suggested that the presence of such beasts as these (perhaps attaining a total diameter of many miles in the case of Jupiter's postulated counterparts to our ocean's fish hunting animals) may actually explain the frequent occurrence over the planet of\ clearly perceived areas of red colouration.

They hoped that this prospect could be investigated by the close-up cameras that would be focused upon Jupiter by the United States' Mariner 11 and 12 space probes following their launch later in 1977, but no evidence to support the existence of these (or any other) types of Jovian life form was acquired. Even so, the ideas of Salpeter and Sagan remain plausible relative to the likely nature of life on that huge world, should life truly exist there.

Yet what of Earth? Earth is not a gas planet, but notwithstanding this it does possess a notable atmosphere. Could such creatures exist here? Momentarily dipping once more into the realms of science-fiction, this concept was the basis of a compelling short story by Sir Arthur Conan-Doyle, entitled 'The Horror of the Heights' (Strand Magazine, November 1913). Predictably, the upper reaches of our skies were found to be inhabited by all manner of gigantic, wholly aerial creatures, of many different types, including, (of course), a stupendous form of sky-jellyfish. However, the short story's hero, a brave but inquisitive aeroplane pilot, also encountered huge varieties of aerial serpent, not dissimilar in appearance from some depictions of the fabled Chinese dragon. To quote Doyle's fictitious flyer:

'But soon my attention was drawn to a new phenomenon - the serpents of the outer air. These were long, thin, fantastic coils of vapour-like material, which turned and twisted with great speed, flying round and around at such a pace that the eyes could hardly follow them. Some of these ghost-like creatures were twenty or thirty feet long, but it was difficult to tell their girth, for their outline was so hazy that it seemed to fade away into the air around them. These air-snakes were of a very light grey or smoke colour, with some darker lines within, which gave the impression of a definite organism'.

An imaginative piece of science-fiction; but in a striking parallel with the science-fact sequel to Arthur C.Clarke's 'A Meeting With Medusa', only a few years passed by before a remarkable report was published in which its author claimed that one of his correspondents, a genuine air pilot, had actually encountered a very similar creature in reality!

This report, whose author referred to himself only as 'A Philosophical Aviator', appeared in the Occult Review (December 1917), and more recently was discussed in detail by Nigel Watson in a fascinating article on airborne anomalies (Fortean Times, winter 1982). The correspondent from whom the philosophical aviator had obtained his astounding account was described by him an being an experienced (but un-named) World War 1 airpilot, who had allegedly confronted at considerable altitude a weird apparition that he likened to a colourful dragonesque creature, floating through the air towards him at an appreciable speed - an unnerving event that, (not surprisingly) had persuaded him to descend to earth at once! As Watson suggests, he was most probably suffering from oxygen deficiency, a hazard when flying at great heights, and capable of engendering a wide range of optical hallucinations.

Needless to say, the conspicuous absence of names for both the airpilot and the 'philosophical *aviator*' unavoidably turns thoughts towards the possibility that the entire report was nothing but a hoax (perhaps even inspired by the Doyle story), but even if this is true, how can we explain the many other reputed sightings of sky-serpents that have been reported over the years, and from numerous localities worldwide?

In *'Curious Encounters'* (1985), Fortean researcher Loren Coleman documented several such cases, which make strange reading indeed. Whatever a certain farmer in Bonham, Texas, expected to see when he looked up at the sky while working on his farm one day in June 1873, it certainly was not the gigantic, yellow-striped serpent, writhing and thrusting, that floated overhead! A hissing sky-snake was also reported during May 1888 in South Carolina's Darlington County, a Scandinavian equivalent was spied over southern Norway and Denmark in May 1935, and nine months later another one appeared over Brazil's Cruz Alta. Nor is Britain immune to overhead ophidians - the Devon town of Bideford was brightly illuminated by a twisting sky-serpent for six minutes on 5th December 1762.

What are we to make of these entities from the ether? Terrors in tangible form? Or could they be monsters of a quite different meterological nature?

Meterological Monsters.

There is no doubt that many purported sightings on record of aerial serpents and dragons (not to mention fire-drakes and other dramatic flame-emitting fauna) simply involves unexpected or misidentified observations of aurorae and other meterological phenomena.

An early example of this can be found in *'Contemplation of Mysteries'*, published in England during the Elizabethan Period. Apparently in the year 1532 *'flying dragons'* were seen in several different countries and were recorded in the above book as ... *'flying by flocks or companies in the ayre, having swines' snowtes; and sometimes were there seene foure hundred flying togethir'*. The *'dragons'* themselves were described as follows:

'The flying dragon is when a fume kindled apeereth bended, and is in the middle wrythed like the belly of a dragon, but in the fore part, for the narrownesse, it representeth the figure of the neck, from whence the sparkes are breathed or forced forth with the same breathing'.

The meterological nature of these *'creatures'* was also readily exposed in a rather later tome of this type, in which the writer Blout stated:

'there is a fire sometimes seen flying in the night like a dragon; it is called a fire-drake. Common people think it is a spirit that keeps some treasure hid, but philosophers affirm it to be a great unequal exhalation inflamed between two clouds - the one hot, the other cold (which is the reason why it smokes), the middle part whereof, according to the proportion of the hot cloud, being greater than the rest, makes it seem like a belly, and both ends like a head and a tail'.

And the manner in which the shape of the 'dragon' is created was succinctly described in *'The World of Wonders'* (1882) edited by A.Taffs:

'When vapours of an inflammable kind collected in the air and ascended to a cold region, the vehement agitation thereby produced induced a flame. The highest part, being more subtle, assumed the singular form of what was presumed to be the dragon's neck, and then, having been made crooked by the repulse it received, formed the dragon's belly, while the hind part, turned upwards by the force of the same collision, represented the monsters tail. Then, with impetuous motion, it fled through the heavens - all ablaze, as it were - striking deadly terror into the hearts of the ignorant and superstitious'.

Snakes with wings.

In addition to the sky-serpents and fire-drakes mentioned above are some decidedly bizarre accounts of snakes with wings. Such creatures occur in ancient legends and folktales from most parts of the world., but there are also some quite recent reports featuring winged snakes of the diconcertingly corporeal kind. Nor do these unlikely beasts appear to be restricted to some extremely remote land only rarely penetrated even by the most intrepid of Westerners. On the contrary, one of the strangest specimens made its public debut in the zoologically-undistinguished locality of West London!

In the 20th April 1798 issue of a long-vanished periodical called 'The Gentleman's Magazine', a semi-anonymous correspondent signing himself only as 'SB' contributed the following remarkable report:

'In the beginning of the month of August, 1776, a phenomenon was seen in a parish a few miles west of London which much excited the curiosity of the few persons that were so fortunate as to behold it. The strange object was of the serpent kind: its size that of the largest common snake (in Britain, this is the grass snake Natrix natrix, averaging 30 inches long, the British record-holder is a female measuring five feet ten inches, from South Wales in 1887), and as well as it could be discovered from so transient a view of it, resembled it by a kind of grey mottled skin. The head of this extraordinary animal appeared about the same size as a small woman's hand. It had a pair of short wings very forward on the body, near its head; and the length of the whole body was about two feet. Its flight was very gentle; it seemed too heavy to fly either fast or high, and its manner of flying was not in an horizontal attitude, but with its head considerably higher than the tail, so that it seemed continually labouring to ascend without ever being able to raise itself much higher than seven or eight feet from the ground'.

Clearly not the most aerially efficient of flyers, and wholly unrecognisable as any species known to science at that time (or since). True, certain species of insect (notably some of the dipterans or two-winged flies, such as the asilids, or robber flies) do maintain their heads well above their bodies while in flight, sometimes adopting an almost vertical posture - but these insects are rarely above an inch in length, not two feet! Damselflies, those delicate relatives of robust dragonflies, possess such a tiny thorax (the portion of an insect's body that bears the wings) in comparison to their very lengthy, elongated abdomen, that they could perhaps be mistaken for an airborne snake bearing wings just behind its head by someone with an exceedingly scant knowledge of natural history, but even these are less than six inches long,

An Asilid or Robber Fly

A typical damselfly

' and unlike dipterans they possess four wings, not two.

If this had been an isolated case, it could be discounted as at best a curiosity with a pleasantly cryptic, but undoubtedly commonplace explanation at its core, and at worse a bizarre hoax. However, this was not the end of the matter.

As recalled in a second account within *The Gentlemans Magazine*, this time by the pseudonymous 'J.R', the same (or a very similar) entity had been seen by a friend in an unspecified locality between Hammersmith and Hyde Park Corner on 15th June 1797 at 10.30 p.m. According to his friends:

'the body was of a dark colour, about the thickness of the lower part of a man's arm, about two feet long. The wings were very short, and placed near the head. The head was raised above the body. It was not seven or eight feet above the ground. Being an animal of such uncommon description, I was particular in noticing the day of the month, and likewise being the day preceeding a most dreadful storm of thunder and lightning'.

Extraordinary though it may seem, even the winged wonder of Hyde Park pales into insignificance when compared with the resplendent poultry predators that once allegedly thrived near Penllyne Castle, Glamorgan, and which reportedly occurred in such numbers up until as recently as the last century that they were routinely shot by irate farmers as vermin!

Their supposed former existence was uncovered by Marie Trevelyan while researching for her fascinating book *'Folk and Folk Stories of Wales'* (1909), in which she included a detailed description given to her by a local eyewitness who died in about 1900:

'The woods round Penllyne Castle, Glamorgan, had the reputation of being frequented by winged serpents, and these were the terror of old and young alike. An aged inhabitant of Penyline, who died a few years ago, said that in his boyhood the winged serpents were described as very beautiful. They were coiled when in repose, and "looked as if they were covered with jewels of all sorts. Some of them had crests (a sex linked characteristic, perhaps?), sparkling with all the colours of the rainbow". When disturbed, they glided swiftly, "sparkling all over", to their hiding place. When angry, they "flew over people's heads with outspread wings bright, and sometimes with eyes too, like the feathers in a peacock's tail". He said it was "no old story invented to frighten children", but a real fact. His father and uncle had killed some of them, for they were "as bad as foxes for poultry". The old man attributed the extinction of the winged serpents to the fact that they were "terrors in the farmyards and coverts"'.

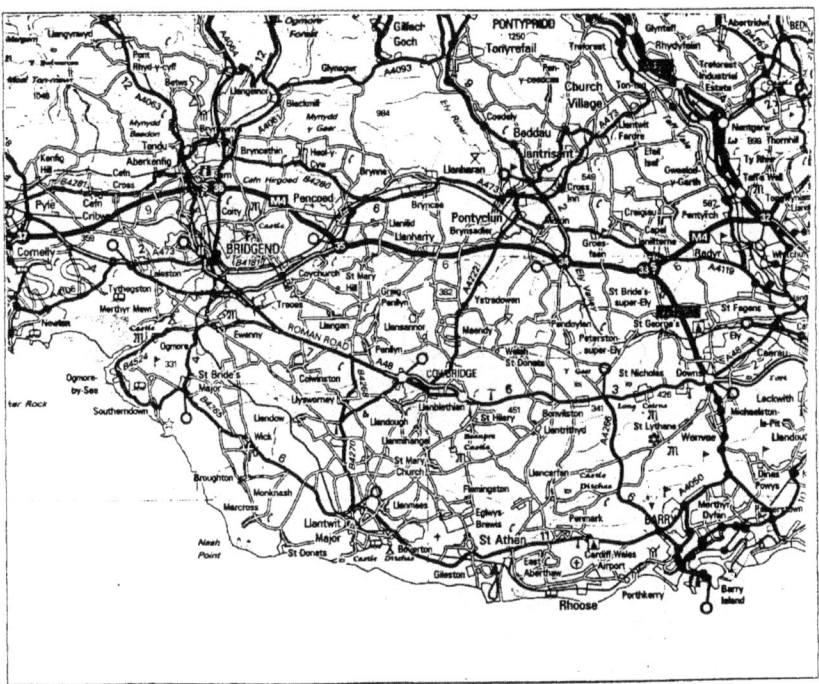

'The district of Glamorgan Pennlyne Castle'

As with the London winged snake, this was not the only such account that came to light.

According to Trevelyan:

'An old woman, whose parents in her early childhood took her to visit Penmark Place, Glamorgan, said she often heard the people talking about the ravages of the winged serpents that neighbourhood. She described them in the same way as the man of Penllyne. There was a 'king and queen' of winged serpents, she said, in the woods around Bewper. The old people in her early days said that wherever winged serpents were to be seen "there was sure to be buried money or something of value" near at hand. Her grandfather told her of an encounter with a winged serpent in the woods near Porthkerry Park, not far from Penmark. He and his brother "made up their minds to catch one, and watched a whole day for the serpent to rise. Then they shot at it, and the creature fell wounded, only to rise and attack my uncle, beating him about the head with its wings. She said a fierce fight ensued between the men and the serpent, which was at last killed. She had seen its skin and feathers, but after the grandfather's death they were thrown away. That serpent was as notorious 'as any fox' in the farmyards and coverts around Penmark'.

On first inspection, the above report seems just too good to be true, many of its principle elements more consistent with a noble legend of valour than a description of an encounter with a member of the local fauna! Dragonesque serpents headed by a king and a queen, the pursuit of one such beast by two valiant hunters, followed by an epic battle in which the accursed creature was ultimately slain and its remains retained by one of its dispatchees as a glorious trophy - even the longstanding association of dragons with hidden treasure is included. As for the creatures themselves, it hardly need be said that winged, plume-bearing serpents comprise a blatant zoological impossibility.

Yet amid all of this apparent fantasy, residing uncomfortably within what could otherwise be discounted without further ado as a charming but fictitious folk story, are two very curious facets - one unexpectedly far-reaching, the other incongruously prosaic.

The first of these is the corroboration that the womans' statement provides for the Penllyne man's account concerning the former existence in Glamorgan of these creatures. That is to say, two people from quite separate localities independently recollected the onetime occurence of winged serpents with plume-adorned bodies in this region of Wales.

The second, and even more intriguing facet, again linking the two independent accounts, is the almost absurdly mundane attitude of the local farmers to those astonishing creatures, whose taste for their chickens evidently attracted far greater attention than their flamboyant appearance - an attitude so inherent that in the eyes of the farmers they were nothing more than troublesome vermin, deserving of no better treatment than would be meted out to foxes or rats!

Indeed, like many other zoologically exotic but agriculturally antagonistic creatures on record - from North America's extinct Carolina parakeet (whose crime was a predeliction for the produce of apple orchards) to Tasmania's striped marsupial wolf (accused of large-scale sheep slaughter, but for which it was never proven to be responsible) - they were relentlessly persecuted until finally they were totally exterminated.

Their taste for poultry also rules out an identity that seems almost as odd as the creatures themselves. As I noted in my book 'Dragons - A Natural History' (1995) when dealing with these Welsh winged wonders, it has been suggested that multicoloured serpents with feathered wings sighted in the Vale of Edeyrnion in 1812 may have been cock pheasants,

'Ring Necked Pheasant'

which were not familiar there at the time. However, pheasants do not devour chickens, and it seems rather unlikely that a pheasant could be mistaken for a flying snake.

Yet assuming that the woman's account was a truthful one, the mystery of Glamorgan's aerial chicken killers, a profoundly curious fusion of the mythological and the matter-of-fact, might nonetheless have been solved - if only her grandfather's relatives had not been so eager to discard the skin and feathers of the specimen that he had supposedly killed, the only tangible evidence for their alleged reality. After all, it's not every unwanted souvenir that could have conceivably comprised the relics of a bona fide feather-adorned, jewel-spangled serpent with wings!

In more recent times, another flying snake has been hitting the cryptozoological headlines - and for good reason. Hailing from Namibia, including among its investigators no less a wildlife authority than coelecanth discoverer Marjorie Courtenay-Latimer, and not content with possessing what eyewitnesses liken to an extensible membrane on each side of its neck region, this airborne python-like anomaly is also said to bear a bright light on top of its head, as well as a pair of curved horns! Horned snakes are not unknown, and the 'light' may be nothing more remarkable than a highly-reflective scale or a cluster of scales - always assuming, of course that this feature is genuine and not the product of imagination, confusion with some natural phenomenon like swarms of fireflies, or an optical illusion.

Also worthy of note, however, is that Namibia is a former German colony, and German mythology includes stories of a fabulous lightning snake. It would not be the first time that a legend from one culture has been incorporated into the mythology of another. In 1995, a South African television company screened a documentary investigating Namibia's reclusive flying serpent, so perhaps the interest engendered will lead to further studies.

EDITORS NOTE: CFZ Researcher, Richard Muirhead has been investigating the giant flying snake of Namibia and the results of his investigation are printed elsewhere in this book.

Are UFOs critters?

Over the years, countless identities have been proposed to explain the appearance of UFOs - including such familiar contenders as extra-terrestrial spacecraft, various metereological phenomena, misidentified aircraft, and all manner of optical illusions - but in 1978 an entirely different and certainly very novel identity received attention, with the publication of 'Sky Creatures'. Written by Trevor James Constable, this highly intriguing book suggested that many reported UFOs were in reality living creatures - highly-modified, native life-forms specialised for an exclusively airborne existence within our planet's atmosphere. (And therefore comparable to the Jovian balloon-like beasts postulated by Salpeter and Sagan).

According to Constable, Earth's sky-creatures (which he has christened 'critters') possess a mica- or metal-like outer body surface, are unicellular in structure and reminiscent in form to amoebae, and exhibit a great range in sizes - from lengths of half a mile down to just a few inches. He also claimed that these 'critters' generally vibrate in the infra-red portion of the electromagnetic radiation spectrum (and hence cannot be detected by human eyes), but on occasion are somehow able to become visible, and even to be photographed - resulting in additions to the ever-expanding files of UFO reports. In his book, Constable even included photos that he has taken of alleged 'critters' during their visible phase; to date, these photos have never been explained - but they have not been denounced as fakes either.

Although Constable's notion of undiscovered aerial organisms as UFOs is, without question, a highly unusual one, it is neither the first nor the last time that such an idea has been voiced, as outlined recently in an excellent review of this subject by veteran ufologist Jerome Clark (Fate, April 1991). In the early 1960s for example, American inventor John M.Cage likened sightings of UFOs trailing aeroplanes to incidents involving dolphins following ships or other seaborne vessels, and he suggested that such UFOs were sentient life-forms charged with (and feeding upon) 'negative electricity' *(Fate, September 1962)*.

A decade earlier, Countess Zoe Wassilko-Serecki had postulated that Earth's upper atmosphere was inhabited by huge bladder-like beasts, subsisting upon energy in order to sustain their glowing, shape-shifting bodies, composed principally of energy but enveloping a central core of solid matter *(American Astrology, September 1955)*. According to the Countess, these enormous, but relatively incorporeal entities, (again recalling the Jovian 'balloon beasts' and Constable's 'critters') are spherical when stationary, but assume cigar like shapes when moving; it is interesting to note that numerous UFO reports since then have indeed featured immense, cigar-shaped objects.

More recently, Richard Toronto obtained a photo of a UFO that closely compares with those of Constable *(Fate, August 1990)*. And bringing the subject of 'living UFOs' right up to date is Andrew Collins's book *'The Circlemakers'* (1992), reviving Constable's 'critter' theory by proffering the contact of such creatures with our planet's surface as part of the explanation for that most contentious of current enigmas - the origin of crop circles.

Electrifying Entomology.

A very different concept of 'living UFOs' from those discussed above has been offered by various insect researchers. In the 1960s, amateur scientist Norton Novitt from Denver, Colorado became interested in the possibility that certain UFO sightings involved swarms of insects that had somehow been rendered luminous. This idea stemmed from a sighting that he had made of two glowing ants in flight one day, their apparent luminosity actually comprising reflected sunlight. Some species of ant engage in mass nuptial flights at certain times of the year, and as these mating swarms can contain several million insects they often attain a very considerable size - large enough to resemble glowing orbs in the sky if there is sufficient sunlight to bounce back to earth from the swarms. Even so, luminous UFO sightings made at night could not be explained by this theory - or could they?

As described by Robert Chapman in his book *Unidentified Flying Objects'* (1968), Novitt wondered whether it was conceivable that flying ants could generate their own luminosity (i.e as distinct from merely reflecting rays of sunlight). To pursue this piquant line of speculation, he attached some winged ants to a ping-pong ball, which in turn was connected by a thin wire to a static generator placed in a darkened room - and, sure enough, when the generator was set in motion the ants' bodies began to glow brightly. Although certainly interesting, such an experiment may appear rather futile at first, because, (as Chapman drily commented in his own coverage of Novitt's researches) in the natural world ants are not normally attached to generators! However, it just so happens, that nuptial flights of ants often take to the air shortly after thunderstorms - weather conditions that give rise to very strong atmospheric electric fields. Under such conditions, it is quite likely that the swarms would indeed glow, and with a light strong enough to be easily observable at night. So perhaps there are some UFO reports on record that were inspired by swarms of flying ants after all.

A few may have featured swarms of moths too. In the late 1970's, insect behaviouralists Drs Philip Callahan and R.W.Mankin from the United States provided independent support for Novitt's findings by revealing that light can be generated by placing specimens of Canada's spruce budworm moth *(Choristoneura fumiferana)* in electric fields - confirming that during those weather conditions when the air is heavily charged with electricity, insects are capable of emitting light. Of course, the amount emitted by each insect would be minute, but as migrating swarms of spruce budworm moths can measure up to sixty miles long and fifteen miles wide, the total amount of light emitted per swarm would be of very appreciable magnitude - more than enough, in fact, to mimic a glowing UFO. As Callahan and Mankin acknowledged, however, it is noticeable that a number of UFO sightings of this type that they have analysed occurred at times when mass migration swarms of this species of moth would be expected *(Applied Optics, 1978)*.

Bones of the Thunder-Horse.

Probably the most dramatic of all cases involving the alleged existence of sky-dwelling creatures is that of the Amerindian thunder horse - for this was based on more than just hearsay. When challenged by scientists during the 1800s to produce proof of the thunder horse's existence, the Sioux Indians of Nebraska and South Dakota were more than willing (and able) to comply - presenting them with gigantic bones that could not be identified with any species of animal known to science!

The thunder horse is undoubtedly one of the most spectacular beasts from the annals of native American Indian mythology - an enormous, terrifying creature that leaps down from the skies to earth during violent storms. According to those legends, thunder is the sound that its hooves produce upon impact with the ground, and it also uses its hooves to slay bison before returning skyward. Supporting their claims, the Sioux were aware that during heavy rainstorms, immense bones were often discovered washed up out of the ground, which they naturally assumed to be the earthbound remains of the thunder horse.

Greatly intrigued by all of this, in the 1870s the celebrated American palaentologist Prof. Othniel Charles Marsh examined some of these salvaged thunder horse relics, and recognised that they were, in reality, the fossilised bones of some long extinct species of giant mammal. Continuing his studies, Marsh deduced that the beast was a gigantic form of perissodactyl (odd-toed hooved mammal), seemingly most closely related to horses, but which had borne upon its nasal bones a massive Y-shaped structure most nearly resembling in composition the specialised horns (bony, skin-covered, non-deciduous, known as ossicones), of the giraffe and okapi.

Brontotherium...the original 'thunder horse'

Now shown to have existed roughly 35 million years ago, during the Oligocene epoch, it is just one of an outstandingly diverse array of species referred to today as titanotheres. Moreover, its intimately-associated thunder horse myth inspired Marsh to christen this particular beast Brontotherium - 'thunder beast'.

Final thoughts.

It can be seen from this varied selection of examples demonstrating the coupling of the sky-beast notion with an emphatically heterogeneous parade of perplexing phenomena that the concept of exclusively aerial life-forms inhabiting our skies and clouds (not to mention those of other planets) is one that has captivated humanity from the very earliest times, and continues to do so even today. In fact, one would be hard-pressed to define the greater mystery - whether the atmosphere embracing our world does indeed harbour strange, spectacular fauna still to be discovered by science, or why the possibility that such creatures exist, holds such an enduring, compelling fascination for us!

Perhaps the simplest solution would be to echo the opinion of the eminent British geneticist Prof. J.B.S.Haldane. As expressed in his 'Possible Worlds' (1927):

> 'Now, my suspicion is that the universe is not only queerer than we suppose, but queerer than we can suppose....I suspect that there are more things in heaven and earth than are dreamed of in any philosophy. That is the reason why I have no philosophy myself, and must be my excuse for dreaming'.

*****

Editor's Note: This article first appeared in a 1993 issue of SCAN News, albeit in a much truncated form. The full text of Dr. Shuker's article has, however never been published before.

*****

THE NNIDNIDIFICATION OF NESS

by Tony 'Doc' Shiels

(Although Tony is a well known figure in the fortean world, most of his published writings have been either on the subject of stage magic or surrealism. This piece, which discusses several of the concepts explored in his book 'Monstrum', was written several years before, and originally appeared in the first, and to date only, issue of 'Nnidnid-Surreality' magazine in 1986. It would therefore have evaded the attention of most cryptozoologists. It has been reprinted by kind permission of the author.).

At 4 p.m on May 21st 1977, I photographed the Loch Ness Monster. With the aid of shamanic surrealism and nnidnidiomancy I invoked the beastie.

A sigil helped. This is it..the nnidnidiogram.

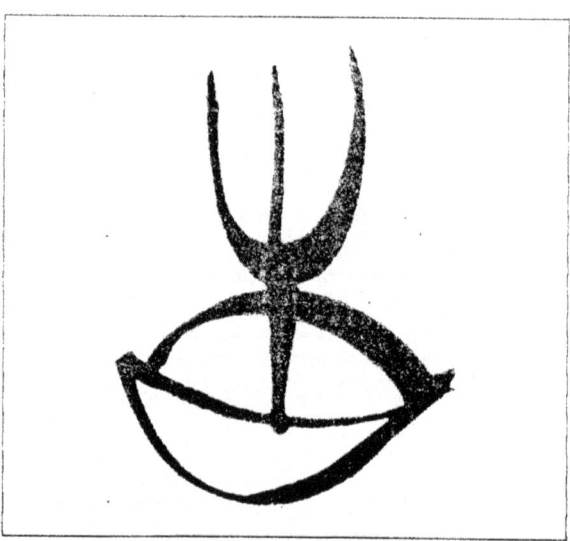

It is an upstanding, three-tined fork in a soup bowl; eye, crescent moon and trident; bum and chamberpot; an awesome insect; penis and vagina; a portable TV set; a gun sight; a rare plant in a plant pot; somebody's something reflected in a round mirror; a very thin person, arms raised in a bathtub. It is a monsterbaitory symbol. It works. It attracts aquatic monsters like a picnic attracts wasps. It works especially well if I paint it on the belly of a naked witch, just above the mons veneris. It works even better if painted on the bellies of, say, three or nine naked witches. This, perhaps, is because the Loch Ness Monster is quite obviously ithyphallic, like so many things.

Often, when on a monster hunt, I draw my nnidnidiogram in the creamy froth of a pint of good draught Guinness (guinnage?). Ireland's famous stout is full of dark humour. Its name, GU-IN-NESS, says it all. The GU means Great Unknown or Grotesquely Upstanding. Such messages are there to be read and understood.....and acted upon.

Objective chance has a role to play.

Levitation is part of the trick when raising cryptozoological oddities. Sometimes it is vertical, sometimes horizontal, always impressive. The secret lies in the way one says, 'Up you come', with confidence and authority.

Apart from Scotland's Nessie, I have raised water dragons in Cornwall and Ireland. All these creatures are Celtic cephalopods of an unusual kind, closely related to Dinoteuthis proboscideus .... 'it swoom by the lappits of its mantle' ...... and that cuttlefish in Nosferatu.

The Loch Ness Monster is a perfect *cadavre exquis*, a synthesis of contradictions, a surreal totem beast. In 1933 it came ashore and frightened Mr and Mrs Spicer. He declared that it was 'an abomination'. I think its a kind of elephant squid with a prehensile proboscis. It likes naked witches, and so do I.

The witches swim, wetting themselves, in order to attract the monsters. Dragons have always been attracted by women, its a well known fact. I only have to mention Andromeda ad a good historical example. Then, of course, there was Anne Darrow. Its the way things are. That's nature for you. Perhaps this tells us something about the origin of mermaids, perhaps not.

Some people have suggested that aquatic monsters are phantoms, psychic holograms, paraphysical entities, rather than organic animals. They may be right ..... *'les fantomes vinrent a sa recontre'* ..... but the phantoms can be photographed. Taine has said that *'perception is a true hallucination'*. I photograph my hypnopompic visions. Perceptual functioning is variable, and perceptions are influenced by desrires, conscious and unconscious. Witnesses of the Loch Ness Monster tend to match Nessie as she is perceived with Nessie as she is understood to exist. Some parents tell their children that there are no such things as ghosts ..... or witches.

In the sixth century, the Pictish King Bude lived by Loch Ness. The Picts knew Nessie well, and they portrayed her in their petroglyphs, alongside naturalistic renderings of bulls, eagles, salmon, deer and other 'common' creatures. Nessie is known as the *'Swimming Elephant'* or the *'Pictish Beast'*. The beast has coiled tentacles and a long, thin proboscis curving over its back. Ness means 'nose' and a 'nose' can be a 'proboscis'. Nessie is a female name. Nessie is an ithyphallic androgynous shape shifting mollusc, neither a snark or a boojum. Imagine its song as it rides the waves.

Molluscs are often eaten for their aphrodisiac qualities. Guinness also has a certain reputation in these erogenous zones. I often enjoy a plate of stir fried cuttlefish washed down with a few jars of stout.

The camera is a magical device. My camera bears the nnidnidiogram in invisible ink, visible only in ultraviolet light. This renders the monsters less camera-shy than they would otherwise be. This, and the witches.

In my work as a wizard, I am known as 'Doc' as in Documents. It means almost noting.

I dream of these monsters, and of the adventurous situations in which they decide to display themselves.

The monsters may be ideas ..... monstrous ideas. The appalling Dali has written: *'To look is to invent'*. And C.E.Montague: *'Rightly to perceive a thing in all the fullness of its qualities, is really to create it'*. Photography fixes moments of perception. It is a cool customer.

Having mentioned the revolting Dali, I should note the fact that in 1983, on Achill Island, Co Mayo, Ireland, just before snapping an albino piast in Keel Lough, I encountered a rotting donkey by the roadside. Golf will not be remarked upon, yet.

A conical hat should be worn by wizards. For too many years, I made the mistake of wearing a tall, stovepipe beaver. It was useful, but not as useful as a pointed black spire of a hat. As for the future, I mean to be wise.

* * * * * * * *

THE RIDDLE OF THE SEA EAGLE
A cryptic raptor from the Westcountry.

by Jonathan Downes.

"The truth of a proposition has nothing to do with its credibility....and vice versa".
Lazarus Long.

(Cryptoornithology is a sadly neglected branch of cryptozoology. The avian anecdotes related here are not strictly cryptozoological, but they are mysterious, and they are certainly fortean. It was written in 1992 for my now abandoned book about the mystery animals of the westcountry. Whereas I have not included any records since 1992, to the best of my knowledge the information contained within each section is still accurate. JD).

One winters evening in 1978 I was driving along the A39 past Hartland and towards Clovelly when I was startled to see a large bird of prey perched on a fence by the side of the road. It appeared to be quite considerably larger than a Buzzard, and my immediate thought was that it was an eagle of some kind. How could it have been? Eagles aren't found in Devonshire.

Or are they?

A surprising number of large birds of prey are uncommon visitors to the western part of these islands. There are even a few reports of such real oddities as the Egyptian Vulture, (two were seen in Somerset in the last Century [1], but these are purely vagrants and they are such uncommon visitors that the chances against them becoming established, either by accident, or as a result of a deliberate introduction, (illegal under the terms of the 1981 Wildlife and Countryside Act), must be astronomical. Amongst the occasional visitors, vagrants and escapees have been four species of eagle.

The Spotted Eagle *(Aquila clanga)*, uncommon even in its natural range across the wide forests and steppes of Eastern Europe and the Transcaucasus, is by far the rarest aquiline visitor to British shores. It is relatively small, when you compare it with its more illustrious and better known relatives, being only 27-9 inches in length and with a maximum wingspan of 67 inches. Even so, it is a striking dark brown bird with speckles of white which give it its common name. [2] [3].

It is known from the region by three records. One bird was shot on Lundy Island by Mr Heaven in 1859 [4] and from two Cornish specimens of 1860 and 1861. The first of these was, paradoxically enough, shot at Hawke Wood. [5] This is not only a ridiculous lexilink but one with important fortean links within the cryptic avifauna of the area. One of the best known British winged zooform phenomena is the 'Owlman of Mawnan', an entity which haunts the churchyard of Mawnan Smith in Cornwall. There have been some other westcountry sightings of similar entities, the best known of which also took place at Mount Hawke. [6] [7].

SPOTTED EAGLE
(Aquila clanga)

Date	Place	Reference.
Winter 1858	Lundy Island	Dymond
4.12.1860	Hawke Wood	Penhallurick
29.10.1861	St Mawgan in Pydar	Penhallurick

Penhallurick N:
Birds of Cornwall and the Isles of Scilly. (1978)
Dymond: *Birds of Lundy* (1980)

Another species, the Tawny Eagle, an african race of the predominantly Eurasian Steppe Eagle *(Aquila rapax)* is known from one record. A single specimen was seen in the Studland area of Dorset in the year 1965/6. Prendergast and Boys (1983) have said that this was probably an escapee from a private collection. [8].

It is quite surprising that whereas African and Indian parakeets are well established in some British locations, and there are a multitude of other records of ludicrously out of place birds (my own records include an Orange Bellied Chloropsis in a garden in Exeter, and also a Timneh African Grey Parrot from a council estate in Harlow, Essex), that there are not more reports of out of place raptors.

The trade magazines are crammed with advertisements offering all shapes and sizes of predatory birds for surprisingly low prices. The small ads column of *'Cage and Aviary Birds'* for the week ending, June 6th 1992, for example, offered Golden Eagles, Eagle Owls, Goshawks and Harris Hawks

Spotted Eagle *(Aquila Clanga)*

Steppe Eagle *(Aquila rapax)*

as well as a multitude of smaller and less exciting raptors for comparatively small sums of money. As Gerald Summers pointed out on a number of occasions during his trilogy of books about captive birds of prey [9] [10] [11], these birds are very adept at escaping from their quarters even after many years of seemingly contented captivity, and it would seem fairly surprising that more have not done so.

The other two species of Eagle which have been recorded from the area have, unlike the proceeding two, at one time or other been breeding residents of the four counties which make up the westcountry. Whilst it is generally agreed that any specimens seen of either species during the past two centuries are purely vagrants, there is a surprising amount of evidence to suggest that the aquiline visits to the south west of England are far more regular than is usually believed. There is therefore the ever open possibility, that, if they have not already done so, there is no reason why isolated pairs of eagles should not occasionally breed, and eventually become as well known members of the avifauna of the west of England as they were three or four centuries ago.

The Golden Eagle *(Aquila chrysaetos)*

The best known British Eagle is the Golden Eagle *(Aquila chrysaetos)*. It is a well known, though relatively uncommon resident of Scotland and the far north of England, and not unsurprisingly a few aberrant individuals have been reported from the South West.

It is an enormous bird, up to 34 inches in length, and with a wingspan of up to 90 inches. It is dark brown in colour with golden yellow feathers on the top and back of its head. It is about one third larger than a buzzard, a bird which it superficially resembles. [12]. The juveniles have a black terminal band and large, white patches on the wings, especially on the underside. Stassny [13] notes that although it is a British resident, it is nomadic in habits. What is actually surprising is that there are so few records from the area with which we are concerned.

Golden Eagle *(Aquila chrysaetos)*

Date	Location	Reference
1698	St Pirran, Cornwall	Penhallurick
pre 1700	Perranporth (several specimens)	Penhallurick
pre 1700	Cornwall	Penhallurick
September 1810	Lanteglos, Cornwall	Penhallurick
1859	Trelowarren & Merthen	Penhallurick
1869	Tregothan	Penhallurick
pre 1871	Lundy	Dymond
pre 1876 Devonshire	Lundy	Transactions Devonshire Association, Vol 8 pp 257
4.2.08	Blandford Forum	Prendergast/Boyes
February 1920's	Exmoor	Hendy
1929	Stover	Torquay Nat Hist Soc 1946-50 p 179
17.1.1941	Hanford House, Dorset	Dorset Nat Hist Soc 1940 pp 128
1982	Exmoor	WMN 14.12.82
1986	Exmoor	Beer
1986/7	Dartmoor	Beer

N. Penhallurick: *Birds of Cornwall and the Isles of Scilly* (1978)
Dymond: *Birds of Lundy* (1980)
Prendergast and Boyes: *A systematic list of the Birds of Dorset* (1983)
H.W.Hendy: *Wild Exmoor* (1930)
Trevor Beer: *The Beast of Exmoor* (1988)

The North Devon naturalist Trevor Beer, who is perhaps best known in cryptozoological circles for his excellent work in researching 'the beast of Exmoor', has reported a well documented Golden Eagle living in the Exmoor area and also presented the possibility that one might be living on Dartmoor. [14] [15]. It would seem likely that the two Exmoor records noted above are the same bird, but whether or not the Dartmoor record is also the same bird must remain a matter for conjecture.

It seems quite likely, that as Trevor Beer has suggested [15], some of the sheep killings attributed to the 'Beast' may be attributed to this stray eagle, a beast who is quite as cryptic and just as voracious a predator as many of the cats who have been suggested as the identity behind the Exmoor 'beast'. It would seem likely, however, that its role as predator of domestic livestock is mostly as a carrion eater. A carcass that has been attacked by an eagle would present some of the characteristics that have been seen in some of the sheep kills reported from the moor.

The first reports of the North Devon Eagle came in 1982 when a local newspaper [16] reported a number of sightings across North Devon. Not surprisingly, as we shall see, many observers assumed that it was a juvenile White Tailed Sea Eagle which had been blown off course in a storm, but Trevor Beer eventually identified it as a Golden Eagle.

Four years later Beer recorded another (?) Eagle [17]. A Parracombe woman had repeatedly seen a *'bird larger than a buzzard'* over the preceeding few years. Enough reports had been gathered for Beer to successfully identify this as another Golden Eagle. It was most usually seen on Exmoor between Ilfracombe and Lynton, and also in the Golden Hill area near Bishops Tawton, just outside Barnstaple which was where Beer saw the bird as it flew by along the side of the hillside in the direction of Torridge country.

In his regular newspaper column Trevor Beer surmised that it had flown off towards Dartmoor where *'it would find seclusion and good pickings'*.

Previous to 1982 the only other Devon records appear to have been a bird that may have been a Golden Eagle that was seen, again on Exmoor on February in the 1920's. [18] The other intriguing record will strike a sympathetic chord with cryptozoologists. In 1929 Colonel Welch Thornton had a Golden eagle living on his estate at Thornton for several weeks. he claimed that he had not liked to publicise this series of sightings for fear of ridicule!

There are a few records from Cornwall. [19]

There are several pre 1700 records to *"...an eagle shot at St. Piran in the Sands (Perranporth"* and *"Eagles have more than once been shot at in the neighbourhood, and one, I think, killed"*, as well as an exceedingly frustrating reference to *"an eagle killed by Mr Trevanion at ----- in 16--"* (sic).

Penhallurick (1978) has suggested that these birds are Golden Eagles rather than White Tailed Sea Eagles. In the absence of any evidence to the contrary, I see no reason to disagree with him. He also collected [20] the only other authenticated cornish record of this species, one that was shot at Tredeske, Lanteglos, near Fowey in mid November 1810.

A luckier bird frequented the woods at Trelowarren and Merthen in 1859 and though fired upon escaped injury. It was never identified with certainty. A similar bird, perhaps the same one, was seen at Tregorthan ten years later. [21]

The only other authenticated Golden Eagle record from the westcountry is from Dorset [22]

"The following is from the Morning Post of February 4th 1908:

An Eagle described as a Golden Eagle has been captured on Lord Wolverton's Irwine Minster Estate, Blandford. The bird, which is in the hands of a local taxidermist, measures six foot and nine inches from tip to tip of the wings, and is in excellent condition".

Historically there has always been a fair amount of confusion between the Golden Eagle and the other quondam UK aquiline resident, the White Tailed Sea Eagle. Examples of this mistaken identity come from all four of the western counties. Penhallurick [23] admits that most of the cornish records of Golden Eagles have actually turned out to be Sea Eagles. Prior to 1838 a White Tailed Sea Eagle was shot at Sherborne in Dorset. It was identified correctly, (after about fifty years on exhibition as a Golden Eagle), by Mansell-Pleydell in 1888. [24]).

The following account was published in 1888 [25]:

" *White Tailed Eagles*

In December it was reported in several newspapers that two Golden eagles had been shot near Tiverton. The second bird was killed on the 27th December on the estate of Mr Norish of Fordlands, Tiverton. It weighed ten pounds. As every ornithologist supposed, these Golden Eagles proved to be White Tailed Eagles, and Mr Cecil Smith of Lydeard House, Taunton, the author of 'The Birds of Somerset', wrote the following letter to The Times with reference to them:

'Sir, I saw in The Times of the second instant a notice of two Golden Eagles shot near Tiverton. I have this day seen one of the birds at the bird stuffers at Taunton, and found it, as I expected, not a Golden Eagle at all but an immature White Tailed Eagle in that state of plumage in which the bird is frequently mistaken for the Golden Eagle. If those who would make this mistake would only look for the tarsus or lower joint of the leg, they would find no difficulty at any age of deciding to which species this bird belongs, as at that joint in the Golden Eagle is feathered down to the toes (sic). In the White Tailed eagle it is bare nearly to the joint above. I will not take up your space by pointing out further distinctions. This alone, as between those two birds is apparent to the most careless observer, and true at all ages. I would only add that I believe that the Golden eagle has never been met with in any of these four counties nor in the Channel Islands. At least, all that I have seen recorded as Golden Eagles, have on inspection, turned out to be immature White Tailed Eagles".

As we have seen, he was wrong in supposing that the Golden Eagle had never been recorded from the area, but was certainly correct in his assumption that the common eagle of the south west was, and indeed always had been, the White Tailed Sea Eagle. Certainly the bird that I saw in 1978, more nearly resembled a Sea Eagle, but this magnificent bird became extinct as a resident of the British Isles in 1913. [26] There is a great deal of historical and archaeological evidence for the historical existence of this beautiful and magnificent species in Devon. Does it still exist in the area? The possibility that there may be a few isolated breeding pairs of Sea Eagles in the area is an unlikely one, but not, as we shall see, impossible.

The White Tailed Sea Eagle [27] [28] *(Haliaetus albicilla)*, is the largest Eagle in Europe. The birds are greyish brown with a white tail and a yellow beak. The juveniles are blackish brown with a dark beak and tail on which the white tail feathers from which the species gets its name gradually appear as the bird reaches adulthood. It was once a resident of the United Kingdom, but the last accepted UK specimen, (an albino), was shot in 1913, and although there have been at least two attempts at reintroduction of the species onto islands in the Hebrides, (the most recent being by far the most successful), its present status within the predatory avifauna of the British Isles is uncertain.

White Tailed Sea Eagle

The tradition that the Sea Eagle had once been resident on Dartmoor, but no longer exists there is an ancient one. In 1914, a year after the last British Bird was supposedly shot, E.A.Elliot wrote: [29]

"The Sea Eagle or Erne used to breed regularly on more than one tor on the moor - there are no less than seven "Eagle Rocks" or Tors on the moor - but the birds that breed there are, without doubt, Buzzards, for to a moorman all large birds are Eagles; but there are one or two exceptions, one being the Dewerstone, where this species undoubtedly bred, and also on the outskirts of the Buckland Woods, where there was an Aery (sic) also".

Penhallurick (1978) [30] made a similar observation regarding an 1831 report of 'Eagles' building a nest on Buryan cliffs. These birds turned out on investigation to be Common Buzzards. There are reasons to suppose that as well as the two sites cited by Elliot in 1914, that Eagle Rock in Holne Chase may have been the site of a genuine eyrie.

My father, the author of a Dictionary of Devonshire Dialect [31], confirmed that the common dialect word for all raptors was 'Eagle', and cited several corroborative references [32].

The symbolic stone eagles which cover the walls of Exeter Cathedral are certainly representations of Sea Eagles rather than Golden eagles [33]). The characteristic wedge shaped tails which can be seen clearly on the carvings are a certain indication of this. Trevor Beer also wrote that it was the Sea Eagle rather than the Golden Eagle that was the most common species in the westcountry during historical times [34]. As far as the carvings on Exeter Cathedral and other buildings in Exeter are concerned, there was so especial heraldic or folkloric significance attached to either individual species of eagle, rather than eagles as a whole, which implies that the carvings were simply of the most familiar species in the area during the fifteenth century, when the Cathedral was built.

The plaster eagles which are so common a feature of Bideford, in North Devon, appear, despite their golden colour, to be merely stylised representations of eagles in general, and do not appear to be of any zoological significance.

There are linguistic clues to the historical occurrence of Sea Eagles (and indeed other large birds of prey) in Devon. Ospreys are known to have nested on Lundy until at least 1838, and in the vicinity of Beer until 1750. They arrived in Beer in April and stayed until August. The Devonshire name for this bird is 'Herriot' and they built their nests at the pinacle of a hill known to this day as Arrats (Herriots) Hill, which is about half a mile south of Beer [35]. The traditional name for the Sea Eagle is 'Erne' (also spelt Erm or Orn). Devon place names such as, Ermington, Earme Head, Ernesettle, Ermington and Erme Pits Hill are in the middle of what would have been, historically, prime sea eagle habitat. [36]). Many of these places are in the most deserted parts of Dartmoor in tracts of wilderness, nowadays only frequented by the most foolhardy, and notable only for the continual military presence. These are one of the more obvious places where, if there are undiscovered populations of large birds of prey in the westcountry today, they would be most likely to live.

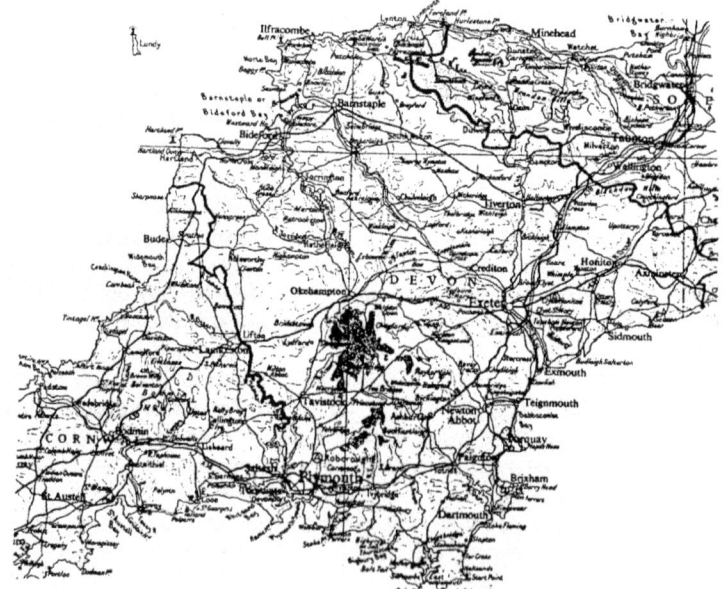

The County of Devonshire.

There is an old country saying that so and so 'can't tell a hawk from a handsaw'. This is used to describe someone who is particularly unobservant. Handsaw is a corruption of 'Earnshaw', another dialect word for the White Tailed Sea Eagle. I was amused to find this old country saying quoted in a science fiction book; 'The Space Family Stone' by Robert Heinlen, a writer who regular readers of mycryptozoological writings, will know that I quote at regular intervals. This is not the first time that characters from his novels have appeared to people the world of cryptoinvestigative theory and methodology!

An 1876 rundown of the Devon avifauna [38], places the White Tailed Sea Eagle well within the predatory avifauna of the region whilst acknowledging its undisputed rarity. Since records began in the mid 19th Century there have been a steady stream of sightings from the South West of England.

Date	Location	Reference
1810-30	Kingsbridge	Trans. Dev. Assoc.1897 p.170
1810-30	Kingsbridge	Trans. Dev. Assoc.1897 p.170
Approx. 1823	Salcombe	Trans. Dev. Assoc.Vol.55 p.117
1834	Devon	Trans. Dev. Assoc.Vol. 8.p.257
1854	Holsworthy	Trans. Dev. Assoc.Vol. 8.p.257
pre 1837	Eddystone	Penhallurick
Pre 1838	Sherborne	Prendergast & Boyes
Pre 1845	Skewjack	Penhallurick
9.11.1845	Kilkhampton	Penhallurick
Autumn 1877	Camborne/Scilly	Penhallurick
1880	Lundy	Dymond
December 1888	Tiverton	Trans. Dev. Assoc. Vol.20 p.40
27.12.1888	Tiverton	Trans. Dev. Assoc. Vol.20 p.40
27.12.1888	Tiverton	Trans. Dev. Assoc. Vol.20 p.40
November 1891	Caerhayes	Penhallurick
December 1893	Bude	Penhallurick
December 1890	N.Cornwall	Penhallurick
November 1897	Lulworth	Dorset Nat.Hist.Soc. 1897.p.203
27.11.1897	Creech Grange	Dorset Nat.Hist.Soc. 1897.p.203
25.11.1899	Stokely	Trans. Dev. Assoc. Vol.32.p.277
Nov 1901	Morwenstow	Penhallurick
Nov 1908	Scilly	Penhallurick
prior to 1923	Salcombe	Trans. Dev. Assoc. Vol.55.p.117
prior to 1923	Start Bay	Trans. Dev. Assoc. Vol.56.p.283
November 1924	Tresco, Scillies	Penhallurick
1932	Dartmouth	Trans. Dev. Assoc. Vol.64.p.219
Feb 1934	Burrator	Devon Bird Report #6 p.245
23.5.34	Lapford	Trans. Dev. Assoc. Vol.66.p.156
11.1.35	Hanford	Dorset Nat.Hist.Soc. 1935 p.97
Jan-Mar 1935	East Dorset	Dorset Nat.Hist.Soc. 1935 p.97
5.11.35	Dartmoor	Devon Bird Watching Report 1938 p.7
prior to 1930	Exmoor	Hendy

Date	Place	Reference.
prior to 1940	Ilfracombe	Ilfracombe Flora and Fauna (1946)
prior to 1940	Ilfracombe	Ilfracombe Flora and Fauna (1946)
prior to 1940	Ilfracombe	Ilfracombe Flora and Fauna (1946)
27.12.41	Wrangaton	Trans. Dev. Assoc. Vol.74 p.136
14.11.41	Eggardon	Trans. Dev. Assoc. 1942
1943	Wigford Down	Trans. Dev. Assoc. Vol.76 p.136
14.1.46	Torquay	Trans. Dev. Assoc. 1946 p.32
Spring 1947	Zennor	Penhallurick
Apr-Dec 1947	Scillies	Penhallurick
June 1948	Zennor	Penhallurick
December 1973	Veryan	Penhallurick
11.8.77	Devon	Trans. Dev. Assoc. 1977 p.24
7.10.77	Devon	Trans. Dev. Assoc. 1977 p.24
Nov 1987	Hartland	CFZ
23.4.91	Monkleigh	CFZ (via Mrs Gribble)

N.Penhallurick: *Birds of Cornwall and the Isles of Scilly* (1978)
Hendy: *Wild Exmoor* (1930)
Prendergast and Boys: *Systematic List of the birds of Dorset* (1983).
Dymond: *Birds of Lundy* (1980)

I have quoted some of the more detailed records in full where they are of specific interest, either because they present evidence to support the theory that the 20th Century status of the Sea Eagle in the South-West, and particularly in Devonshire, may be more solid than has otherwise been supposed, or are of tangental relevance to those of us interested in the chequered history of the British naturalist.

Two particularly obnoxious memoirs of early 20th Century would be naturalists, which are essentially little more than a list of rare birds shot by them or their cronies include 1923 records of White Tailed Eagles from the marshes near Salcombe.[39]

"A few years ago a gunner got a shot at some wigeon off the shore just opposite here one November night and stopping some sent his dog out to retrieve them, when a White Tailed Eagle suddenly swooped down and tried to drown the dog: Of course the gunner had no option but to shoot the bird to save his dog; curious enough, another specimen was shot nearly a hundred years ago in Halwell wood, which is close by".

I have heard that story somewhere before. I think that it was originally published as part of one of the spurious tales of Baron Munchausen but I cannot be sure. It does not sound like a real event, and I feel that there is a definite ring of folklore to it. It feels like what is now known as a foaf-tale (friend-of-a-friend). My gut reaction to this story was confirmed when I found another White Tailed Eagle reference [40], which is so similar as to be practically identical:

"A few years ago a farmers son saw a large bird drop suddenly into (Start) Bay, and come out with a large Perch. He got his gun and shot it - an uncommon bird's usual fate. It proved to be a Sea Eagle. A second example was shot not long after on the estuary as it swooped at a wildfowlers dog which was retrieving some wigeon and would have drowned it if summary proceedings had not been taken".

I think that the event described is such an unlikely one that here we have a piece of local folklore in the making. The story which was taken from another source has been embroidered and has already travelled about fifteen miles. These stories to provide supporting evidence that this rare and wild raptor has been an uncommon but well known sight in South Devon for many years. The long list of slaughtered birds is daunting reading, and indeed the second report continues with this unpleasant vignette:

"In the old coaching days fifty years ago or more the guard just about this spot was handed an Osprey which the keeper had just shot, and he parted with the bird to the bird stuffer for a sixpenny glass of grog".

In the midst of these appallingly callous accounts of avian slaughter it is unusual to find a report dating from either the 19th century or even the first parts of the 20th century which actually condemns the ceaseless slaughter of these unfortunate raptors. Mr W.G.McMurtrie, the manager of Lord Carlingford's estate wrote as follows to a Bristol newspaper in 1888 [41]:

"I note with regret the news contained in a paragraph of Friday, December 30th to the effect that a fine specimen of the Golden Eagle has just been shot at Bagshot by one of Her Majesty's keepers with a spread of fully nine feet. A few days ago there appeared an article on the destruction of Indian birds, an article which contained facts which should have the attention of every natural history society and naturalist desirous of affording protection against the unnecessary slaughter of birds; but it should not be forgotten that charity begins at home and it appears to me far more might, and ought to be done in this respect in our own country. Whenever we are favoured (a not unfrequent occurrence), with a visit from one or more foreign or rare local birds, immediately the person, or persons who have been favoured with a glance at them procure a gun, and under the deceptive title of a naturalist, proceed to riddle (provided the individual can boast of being anything of a shot), the unfortunate birds with lead. Now I am quite aware that in the furthersome of science some destruction must occur; but I am equally as sure that only a small percentage in the way of result is obtained after all of this destruction.

It is possible, if not probable, that if some of these feathered visitors were spared and protected instead of being shot on each visit, in a short time we might be favoured by possessing the species in far greater numbers. At any rate I suppose that we must take it for granted that when a rare bird does happen to get away from our shores after an unaccustomed visit the reception accorded to such a bird is hardly of a nature as to form an encouragement to any other of his race to follow his example and pay a visit to our shores.

I would refer back to the case mentioned above; viz the shooting of an eagle by Her Majesty's keeper. The bird has been taken before in Somerset and in other places, and if the keepers were satisfied in their minds that the bird was of the above mentioned species and had reported it duly in a scientific paper, then surely they would have accomplished as much in the name of science as what they have done; viz shot it. This is only one case out of a great many of this nature. Such destruction is nowhere more feelingly expressed than in the words of Longfellow in his 'Slaughter of the Birds'; viz:

'A slaughter to be told in groans not words'."

There is another well attested record of the White Tailed Sea Eagle from Devon at the end of the 19th century. The final result is again depressingly familiar [42].

"November 25th 1899. An immature White Tailed Eagle was shot whilst perched on the roof of a barn at Stokeley. It had been observed for many days before being secured and during that time had found the Ley a happy hunting ground"

The mixture of brutality and anthropomorphism in that paragraph beggars belief but provides a valuable insight into the mind of the Victorian naturalist.

The records or Devon Sea Eagles are not just a catalogue of killings. Some sightings are specifically interesting from a Fortean point of view.

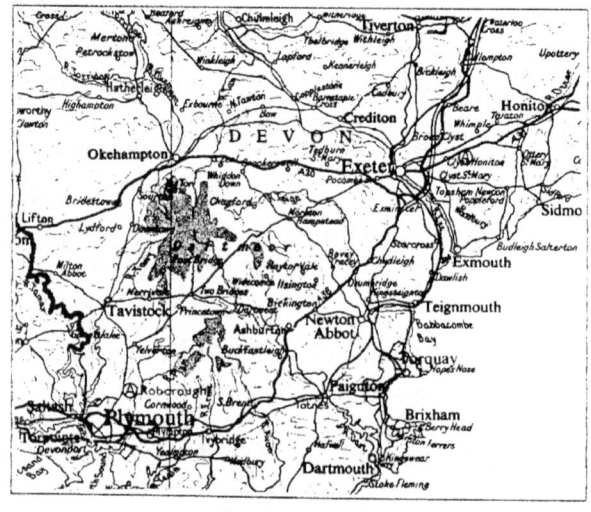

Dartmoor

A well documented series of sightings took place on Dartmoor in 1936. [43]:

"An eagle which proved to be an immature (first or second year) White Tailed, frequented a tract of Dartmoor, east of Princetown during February and March. First reported on February 15th by Miss May of Hexworthy, and subsequently seen by several observers. I was fortunate in securing a very close view of the bird on February 23rd when it rose from a slope behind me and circled round at close range for several minutes. Various plumage characteristics were seen and it may be noted that there was no indication of any white or pale *colouration to the tail. Some interesting skirmishes with ravens were witnessed. Fox Tor Mire and Swincombe reservoir seemed to have a particular attraction for it, but it also ranged into the neighbourhood to the south east and northwards to Laughter Tor. It was seen again on four occasions during a circuit of this area on March 1st. It was reported to have been seen on March 14th for the last time".*

The report was signed G.M.S who wrote again the following year [44]:

"An eagle appeared on the part of the moor frequented by an immature of this species in the early months of 1936, but this time in the unusual month of June. It was seen on at least three occasions and probably was the same bird".

The Fox Tor Mire and Hexworthy areas of Dartmoor are notorious for sightings of strange beasts. The very name Hexworthy has obvious occult connotations, and whilst there are no reasons to suppose that this was anything but an immature Sea Eagle which frequented the same area of Dartmoor on two successive years, the fortean aspects to the geographical location should be considered. One must also ask the question of where the bird disappeared to between March 1936 and June 1937, and what was it that tempted it to return?

H.G.Hurrell, the Devon Naturalist best known for his work with Pine Martens, saw a White Tailed Sea Eagle in the garden of his home at Wrangaton on the 27th December 1941. [45]:

"An eagle mobbed by ravens. Very dark in colour except for some greyish patches on the back, and I think on the upper surface of the tail. The noticeably heron like flight referred to in 'The new handbook of British Birds' seemed to confirm that it was an immature bird of this species".

This confirms a hitherto unwritten law of Fortean Zoology. This law is usually concerned with zooform phenomena, but it seems obvious from the above account that it can also work with flesh and blood creatures:

"Sightings of anything are only made if there is someone to see them".

(I once wrote that the quickest way to destroy any science was to delineate its rules in writing).

This 1941 sighting took place in the garden of the man, who was one of the only people in Devonshire at the time (remember this was at the height of WW2 when every able bodied man between 18 and 41 was liable to conscription into the armed forces, and every older man of fighting age was likely to be still away at the war), who would have understood the significance of what he was seeing.

It is tempting to suppose that somewhere in the nth dimension someone is pushing White Tailed Sea Eagles through psychic window areas so that eminent zoologists can see them, and so that farmhands can shoot them. As scientific logic persuades us that this is unlikely, then the logical inference is that these birds are more common than has otherwise been supposed.

Ten days after I originally wrote the above paragraph I received startling confirmation of my (non) theory.

At 1.30 PM on Wedensday 24th July 1991 the family were on an outing so visit my 'in laws' near Kingsbridge. We were driving along near Loddiswell when an enormous, grey-brown bird of prey swooped in front of my van and was clearly visible for nearly half a minute before disappearing into some woods on the right hand side of the road. It appeared to be about twice the size of a buzzard (which is a bird with which I am very familiar), and although I am only too aware of the danger of laying ones self open to the charge of manipulating the evidence in order to suit the demands of ones theory, neither my wife or I could suggest any identity for the bird we saw other than a White Tailed Sea Eagle.

This sighting took place only days after I had questioned the coincidental sighting of a specimen of this species in 1941 by the man who was probably the most qualified to understand the significance of his sighting. Fifty years later, the same thing happened again. If we could explain coincidences like this (and as I write I can hear the voice of Tony 'Doc' Shiels growling in my head that 'There are NO such things as coincidences!!!'), then we would know a lot more about the nature of space-time reality than we do at the moment.

Because of the peculiar aspects to this sighting I have not included it as a bona fide record of H.albicilla. I am not sure whether it is a bona fide record, a zooform phenomenon, the manifestation of some quasi-jungian wish fulfillment fantasy in some archetypal form, but the para-synchronicitic elements of the event, preclude its inclusion amongstthe purely zoological body of evidence.

The available evidence, even when para fortean elements are excused, has proved that a large number of these huge birds of prey have been seen in the Devon area. We must ask, however, how many other Eagles have visited the area, been ignored, or indeed been misidentified.

Arctic Gyrfalcon

There have been several Devon reports of Gyrfalcons in recent years. Although considerably smaller than any eagle species, it is a magnificent bird. It hails from the icy wastes of the arctic and is thus a more unlikely visitor to these shores than the White Tailed Sea Eagle. One was killed in Instow in 1903,[46], and another was seen off Ballard Head in Dorset on February 5th 1912 [47]. Several were seen in 1972. [48] They were seen at Putford, Foreland Point and on several occasions on Lundy Island. A few years later [48] one was seen in Exminster Marshes. The 1972 report by the Devon Birdwatching and Preservation Society had a footnote added:

"The records have been accepted by the Rarities Commission. At least four Gyrfalcons were seen in southern England, and perhaps more than one bird is involved".

One can make certain interesting deductions by comparing some of these reports with the 1977 Eagle sightings in the Buckfastleigh region [49] which was treated with such manifest disbelief and mistrust that they were only accepted under conditions of stringent scepticism.

In my files I have a number of ridiculously anomalous bird sightings, including one, of a Southern Atlantic Sheathbill inadvertently transported to Plymouth after the Falklands War. [50]. I even have well attested sightings of the Wandering Albatross, a bird only found in the southern hemisphere. 'Doc' Shiels suggests that sightings of Albatross type birds could actually be of the celtic Ean Sidh or fairy bird. [51], but there is no reason to suppose that they are not stray individuals of the southern hemisphere population. The Rarities commission have accepted them as such without any great difficulty.

Why do they, and other *'official'* ornithological bodies, find it so difficult to accept so many modern eagle sightings?

There is an aura of wonder that surrounds the very word 'EAGLE' and there are so many connotations culturally, semantically, etymologically, historically and even spiritually that sightings of these magnificent birds are imbued with connotations absent from sightings of rarer, but less noble birds. There is no doubt, in my mind at least, that the semantically induced scepticism which surrounds so many eagle sightings may well be responsible for the comparative lack of recent reports.

There is no doubt that the White Tailed Sea eagle was the common eagle of Devonshire in times past. There are, however two pieces of evidence to suggest that their current status is more interesting than is generally supposed.

A number of the sightings have referred to birds markedly darker in colouration than would otherwise be expected. This is a series of coincidences which suggests that a specific Devonian variety may have evolved from a genetic pool severely limited by small numbers of individual birds.

Secondly. The sightings have usually been in the areas where one would most expect them to be if the birds were breeding in Devon. Large portions of the craggy wastelands of northern Dartmoor have been the sole preserve of the Army now for many years. The pattern of sightings would suggest that the birds may breed in the relatively unexplored parts of northern Dartmoor, and then fly down to the estuary at the mouth of the Dart to feed. There is also another possible explanation. Anyone who has sailed up the River Dart from Dartmouth to

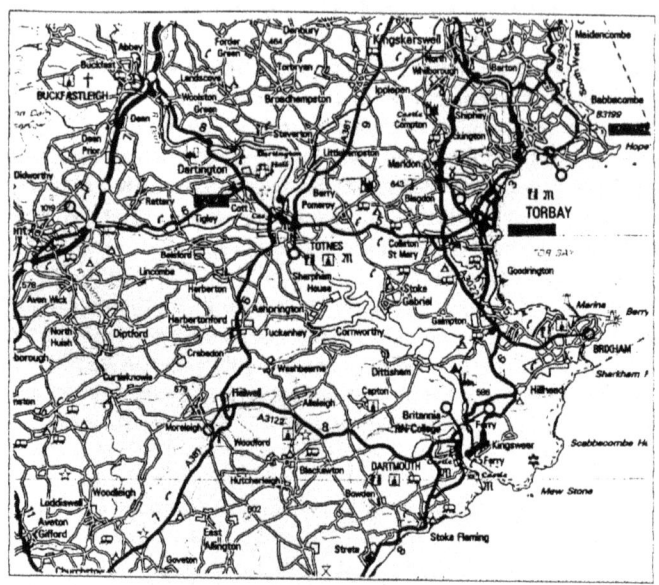

The River Dart, from Dartmouth to Totnes

Totnes must have noticed the dozens of isolated little bays, and the wooded hillsides and inlets along the river where large birds could breed. There is every reason to suppose that if the species has survived as a breeding resident of Devonshire (and I would like to remain neutral on that question) then one of the two areas I have described might be a perfect breeding ground. Recent events in Norway have proved that even such a sizeable bird can be astonishingly cryptic. A 1991 census suggests that there may be twice as many breeding pairs in the country than had otherwise been thought. [52]

The White Tailed Sea Eagle is a widely distributed and ancient species. In common with many other species of raptor it has been persecuted to the brink of extinction across much of its range. By the mid 1980's its European range was confined to the Balkans, parts of the Baltic and the near arctic regions of Scandinavia. Before legal action and scientific enterprise could correct this sad state of affairs a new threat arose from pollution by chemicals and heavy metals. [53]

They are under threat from man throughout their range. Even in the UK where they are strictly protected the vociferous farming lobby, not content with poisoning the environment with chemical pesticides, and injecting livestock with unnatural growth hormones, still trap and poison natural predators, including eagles, and other large raptors on the grounds that they are a menace to sheep.

In 1973 a young bird was seen at Nare Head, Veryan and it provoked the usual hysterical xenophobia from the farming community who expressed a great deal of noisy concern that it might prey upon their hapless sheep, and seriously petitioned the authorities for permission to shoot the bird. Although the authorities would probably not have granted such permission, it seems very likely that had it not merely caught a number of rabbits over a three week period and then disappeared, then some unnamed bucolic sportsman would have taken his shotgun and dispatched it, for the best possible commercial or sporting motives. As we have seen, these birds are extremely unlikely to prey on domestic animals of any kind, and whether or not these birds are still cryptic residents of the wilder parts of the Devon countryside, there is no doubt that they are a regular if uncommon visitor. The big question is, however, that if they are not part of a hitherto overlooked population of Devonian eagles, (and my better judgement suggests that they are not), then where do these birds come from?

A well planned reintroduction programme has been underway for the past twenty years based around the Scottish island of Rhum. It is, of course, not inconceivable that some of the more recent Devonian eagle reports are from this nascent, but now well established breeding colony. The eighty two birds that were used in this breeding experiment were chosen from the Norwegian race of the Sea Eagle, a bird known for NOT being a wanderer, unlike the birds from some of the other races.

The southern Baltic race of these birds is far more prone to wandering far beyond its natural habitat and even now there are a few isolated breeding pairs in what used to be East Germany. If the birds that have been sighted over the years are in fact vagrants they are most likely from this small relict population. There is, however, no discernable pattern of appearances, and although many of the birds seen are juveniles (the immature birds seem more prone to epic wanderings than the adults), but an equal number have not been. [53]

If there is still a relict population of Sea Eagles living in the heart of Dartmoor, it would have more zoological significance than would otherwise have been thought. Love (1988) [53] published what is believed to be the only extant photograph of a living Sea Eagle of the British race. If the species has survived in Devon for the last hundred years or so it would seem, that contrary to scientific belief the British race is not extinct after all. The breeding population that has existed in Scotland for the last fifteen years or so is, after all of the nordic race, and the vagrants which have been reported from various parts of the country are of the Baltic, the Balkan or even the Middle Eastern races. If there is even the smallest of possibilities that the ancestral British race survives it would be in Devon, rather than anywhere else in the country.

Although it is an undeniably attractive possibility, the truth of the matter is, I fear, somewhat more prosaic.

Of the forty seven records I have identified only thirty one of them are detailed enough to provide information for statistical analysis, but the data extrapolated is of considerable interest.

65% of the birds seen were during the period October-January when they would normally be engaged in their pre-mating courtship flights. A number of these birds were, in fact, in pairs, a situation which bears out what we know about their ritualised courtship behaviour. 25% of the birds seen were during the period when they should have been nesting.

The available evidence would suggest that birds of one of the Eastern European populations regularly migrate to the south western counties of Britain to perform their courtship rituals.

Although there is no direct evidence to support this theory, I would suggest that there is every possibility that not only did the species survive in the south-western counties of England for longer than has otherwise been thought, but that they still visit regularly and may even nest occasionally either on the wild and inaccessible crags of northern Dartmoor or in the almost inaccessible parts of the estuary of then River Dart.

Despite my dislike of much of the methodology of the Victorian naturalists, I have always been a vehement supporter of the Victorian fascination with zoological oddities. There are sound zoological reasons for collecting data on such creatures and although they are unjustly ignored by many mainstream naturalists in contemporary zoology at the Centre for Fortean Zoology we are very interested in such things and have many records on our files.

A melanistic buzzard was shot on Dartmoor in 1916 [55]. A melanistic Honey Buzzard was shot on South Dartmoor on the 20th September 1904 sixty four years after a similar bird was shot at Woodleigh woods.[56] I would like to point out that such colour abberations are often the result of a diminishing genetic pool within an isolated population of creatures. As we have seen the last known British specimen of the White Tailed Sea Eagle was an albino. The eagle seen by Hurrel in such mysterious circumstances in 1941 was unusually dark in colour as were some of the Cornish specimens and most markedly the bird seen in Torquay in 1936. If there is an isolated population of eagles eking out a cryptic existence in some hitherto overlooked backwater of the westcountry, (and I would reiterate that this hypothesis is highly unlikely), then they would be likely to, exhibit just these patterns of aberrant colouration.

At the time of writing in mid 1991 the RSPB reports that:

"After many years of persecution and problems caused by pollution the huge White Tailed Eagle is making a determined comeback over northern and central europe".(54).

Even if the suggestions in this paper prove to be unfounded then there is every possibility that this stunning and beautiful raptor may again become a familiar sight for those of us who frequent the wilder parts of the country.

REFERENCES

1. Saunders: *Rare Birds of the British Isles* (1991).
2. Bruton and Singer: *Birds of Britain and Europe* (1970).
3. Stassny: *Birds of Britain and Europe* (1990).
4. Transactions of the Devonshire Association Vol 8 pp 257
5. The West Briton 24.6.1861
6. Bord and Bord: *Alien Animals* (1980).
7. Mawnan-Peller A: *Morgawr-the Monster of Falmouth Bay* (1976).
8. Prendergast and Boys: *Systematic List of the birds of Dorset* (1983).
9. Summers G: *The Lure of the Falcon* (1974)
10. Summers G: *Where Vultures Fly* (1975)
11. Summers G: *Owned by an Eagle* (1976)
12. Bruton and Singer op. cit.

13. Stassny op. cit.
14. Beer Trevor: WMN 14.12.82
15. Beer Trevor: *The Beast of Exmoor* (1988)
16. Beer Trevor (WMN) op. cit.
17. Beer Trevor op. cit.
18. Hendy H.W: *Wild Exmoor* (1930)
19. Penhallurick: *N.Birds of Cornwall and the Isles of Scilly* (1978)
20. Penhallurick N op. cit.
21. The West Briton 24.6.1861.
22. Transactions of the Dorset Natural History and Antiquarian Field Club 1908. p 203.
23. Penhallurick N op. cit.
24. Prendergast and Boyes op. cit.
25. Transactions of the Devonshire Association Vol 8 pp 257
26. Love J.A. RSPB (1981)
27. Bruton and Singer op. cit.
28. Stassny op. cit.
29. Transactions of the Devonshire Association Vol 46 p 497
30. Penhallurick N op. cit.
31. Downes J.T.: *A Dictionary of Devonshire Dialect* (1988)
32. Hare C.E: *Bird Lore* (1952)
33. Transactions of the Devonshire Association 1966.
34. Beer T op. cit.
35. Transactions of the Devonshire Association 1966.
36. A list of Dartmoor place names
37. Hare C.E. op. cit.
38. Transactions of the Devonshire Association Vol 8 p 257
39. Transactions of the Devonshire Association Vol.55. p.117
40. Transactions of the Devonshire Association Vol.56. p.283
41. Transactions of the Devonshire Association Vol.20. p.40
42. Transactions of the Devonshire Association Vol.32 p.277
43. Transactions of the Devonshire Association Vol.69 p.235
44. Devon Bird Watching and preservation Society Reports 1938 p.7
45. Transactions of the Devonshire Association Vol. 35 p.127

46. Transactions of the Devonshire Association
Vol. 35 p.127
47. Proceedings of the Dorset Natural History and
Antiquarian Field Club 1918 p.51
48. Devon Bird Watching and preservation Society reports
1972 p.13
49. Devon Bird Watching and preservation Society reports
1977 p.24
50. WMN 6.10.82
51. Tony 'Doc' Shiels: *Monstrum* (1988)
52. RSPB:*The magic of Birds* (1991) p.9
53. Love J.A:The return of the Sea Eagle (CUP 1983).
54. RSPB:The magic of Birds (1991) p.9
55. Transactions of the Devonshire Association
Vol. 48 p.287
56. Transactions of the Devonshire Association
Vol.39 p.79

WMN = Western Morning News.
E&E = Exeter Express and Echo.
CFZ = Centre for Fortean Zoology.
JD - Jonathan Downes

* * * * * * *

Catmap! A software package to model and study spatial and temporal distributions of Alien Big Cat reports.

Alistair D. Curson

Abstract
A cataloguing system and software package designed to model, spatially and temporally, aspects of Alien Big Cat reports from the British mainland is presented. A preliminary study of reports from the past 35 years has been performed. The trends in the numbers of reports and their distribution are discussed and potential explanations considered.

Introduction
The phenomenon of mystery cats is world-wide with reports dating back for decades[1-4]. Mystery felids may be generally grouped into three categories:

1. Unusual forms of known animals - often the result of rare genotypes. eg. Blue Tigers[2] and the King Cheetah[5,6]
2. Unknown species - eg. the Iriomote cat (discovered in 1965) and possibly the Onza, still awaiting taxonomic classification[7]
3. 'Out-of-place' cats - eg. Jungle cats in Shropshire[8] and pumas captured in Scotland[2].

Reports of mystery cats from the British Isles can be reliably traced back for over two hundred years. Shuker[2] reports of a sighting made at Waverley, Surrey during the 1770s (most probably a lynx). Several sightings have been reported during the early decades of the twentieth century but public interest in the phenomenon really took off with the 'Surrey Puma' in the 1960s.

This project has two main aims:

1. To catalogue British Alien Big Cat sightings in order to generate a readily accessible database of information on these reports
2. To develop a software package to model the spatial and temporal distributions of the reports based upon parameters determined by the researcher (for example annual or seasonal patterns in distribution of reports, distribution of coat colours, distribution of identified species etc.).

It is hoped, ultimately, to use this model to help address a series of questions concerning this phenomenon. Does the model reveal any patterns and what do they suggest? Is the distribution of reports typical of reliable sightings of a wild animal(s) or of random hoaxing? Do they indicate many or few animals? What species are out there and where might they have come from? Are they breeding and can the model tell anything regarding the ecology and behaviour of the cats? Where are the cats living and what are the implications for the British ecology?

Materials and Methods

The catalogue 'UK Felid Historical' (see Fig. 1.) was developed from all reports gathered from literature available to the author[1-3,8-23]. It covers a wide range of reports from the British mainland including 'out-of-place' cats, sightings of unknown felids, strange cries and the discovery of unexplained tracks, droppings and prey carcasses. The catalogue is written in HyperCard v. 2.1 (Claris®) running on a Macintosh LC 475 (Apple Computer, Inc.) home computer.

Catmap! v. 4.2 is also written using HyperCard v. 2.1 (see Figs. 2 and 3). The software programme is written using HyperTalk, HyperCard's scripting (programming) language (very similar to BASIC). Data of interest is read from a plain text file and placed in a memory array. For each independent report the programme takes details of its location (such as longitude and latitude) and calculates the coordinates of the report relative to a map held in the programme's memory (see Fig. 3). This point is then plotted. Locations of other reports from within the operating parameter are plotted to reveal a spatial model. Reports from different parameters may be plotted on different cards (maps) and then the sequence of cards animated so that temporal patterns may be studied.

Fig. 1. Example data card from the HyperCard stack 'UK Felid Historical' - a catalogue of British ABC reports.

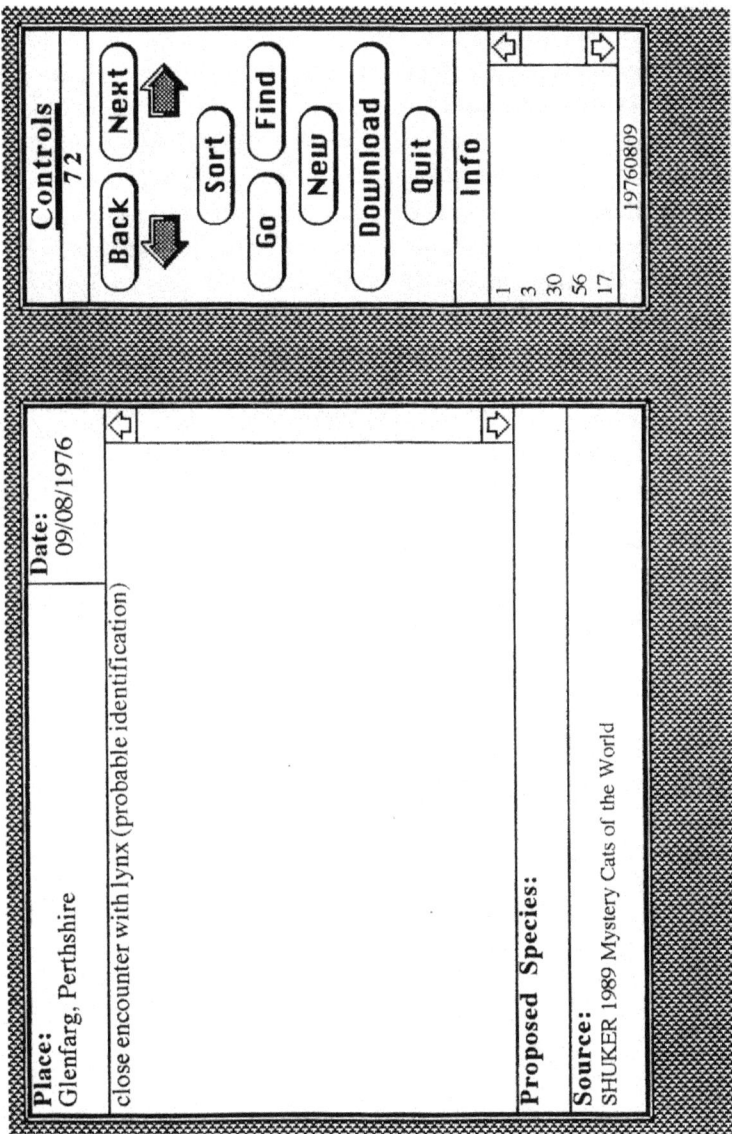

Fig. 2. Title card to *Catmap!* v. 4.2.

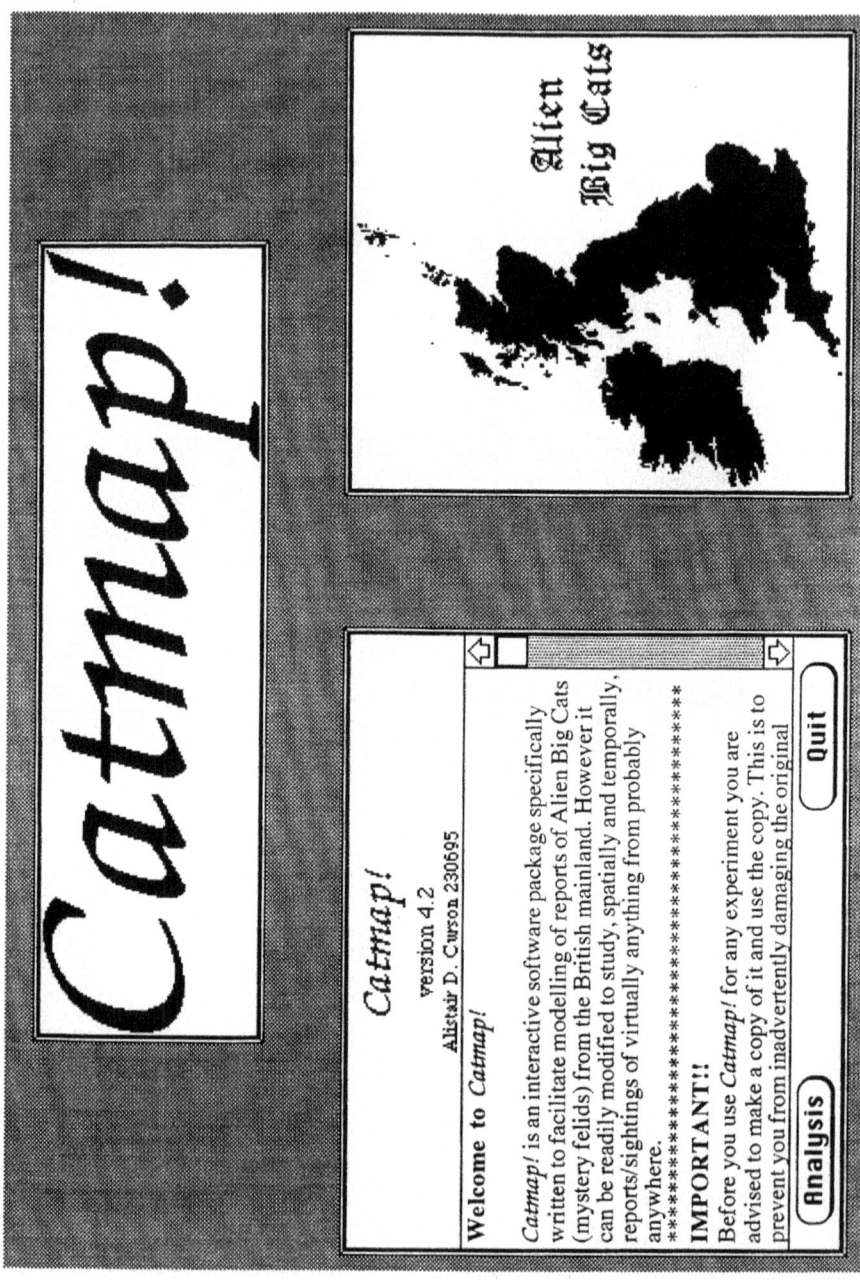

Fig. 3. Example of analysis card of *Catmap!* v. 4.2

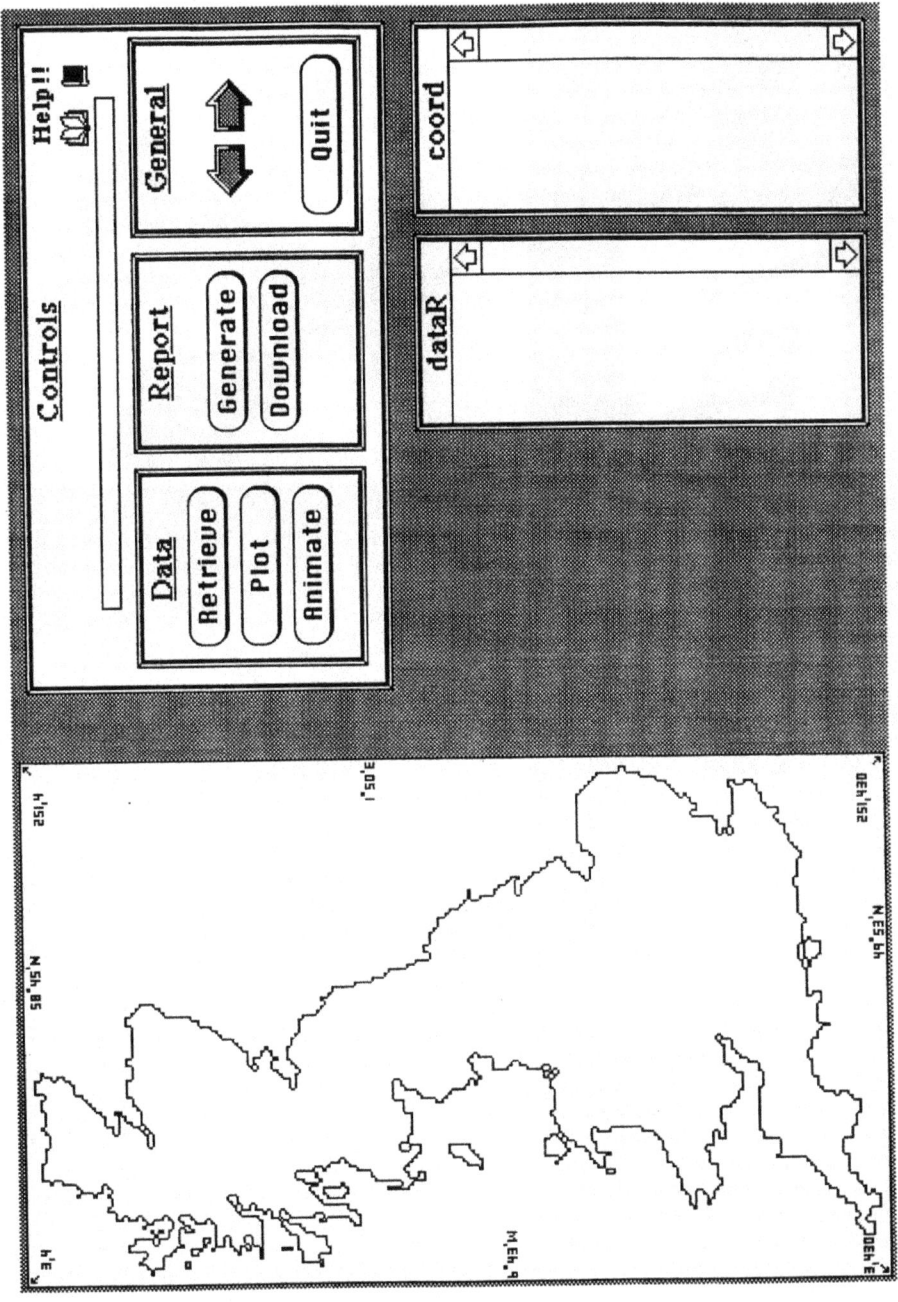

Results
1. Twentieth century reports from mainland Britain
346 reports have been collected for the first 94 years of this century[1-3,8-23]. 299 of the given locations have been found[24-26] and the model calculates 210 independent plots. These are shown in Fig. 4.

The reports are widely distributed over the island, but may be divided into seven loosely defined regions:
      1. The northern Highlands of Scotland
      2. The area surrounding Edinburgh and Glasgow
      3. Tyneside and Yorkshire
      4. The length of the Pennines running into south Wales
      5. Norfolk
      6. London and the Home Counties
      7. The south-western peninsula of England

2. An animation of reports from the 1960s, 1970s, 1980s and the first half of the 1990s
Only six of the reports described in section 1 occurred prior to 1960. Two of these are from Scotland (Inverness and Ullapool), the other four from the south of England (two from Surrey and two from Hampshire). Fig. 5. shows a histogram of the number of reports from each of the four decades under study.

Fig. 4. Twentieth century reports of ABCs from mainland Britain.

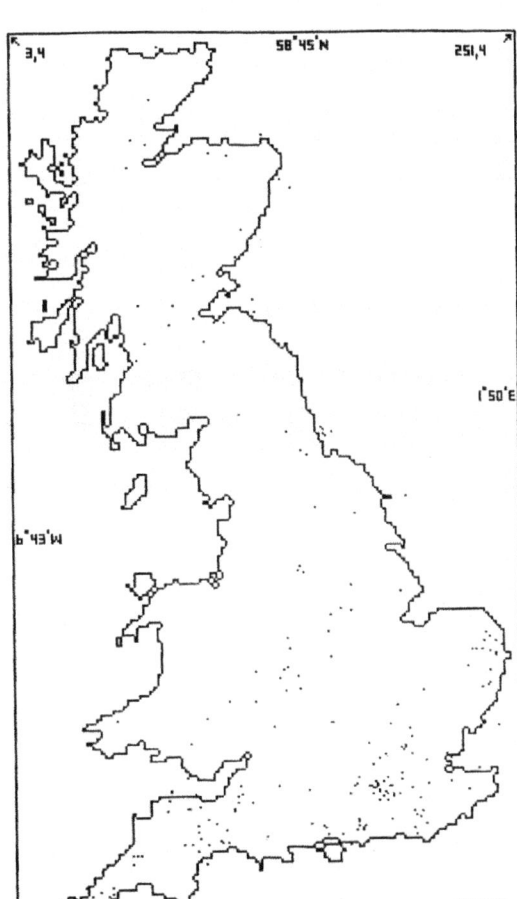

Fig. 5. Frequencies of ABC reports since 1960. For each decade the number of reports is shown along with the corresponding number of locations found and independent plots determined by *Catmap!* v. 4.2.

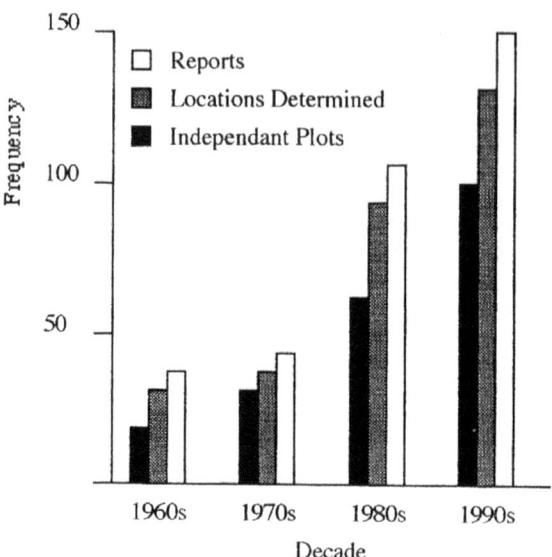

Over the four decades studied, the number of reports remain reasonably constant during the 1960s and 1970s. During the 1980s the number of sightings double, and during the first half of the 1990s there is a further 50% increase. It should be born in mind that the 1990s reports are only those up until the end of 1994.

Each of the four decade report collections was put through *Catmap!* v. 4.2, and the results are presented in Figs. 6 to 9.

All the reports for the 1960s are from the south-east of England and mainly occur to the west and south-west of London. Many of these sightings probably relate to the 'Surrey Puma'.

During the 1970s the number of reports remains unchanged from those of the 1960s however they show a much wider national distribution. The highest densities of reports are from the Edinburgh/Glasgow region, and from the Home Counties. The northern Highlands of Scotland and the south-west of England show a lower density of reports, with a few scattered reports from the Midlands.

The wide national distribution of reports is maintained during the 1980s. It was in 1983 that the 'Beast of Exmoor' hit the headlines with the associated military involvement.

Fig. 6. Reports of ABCs from mainland Britain during the 1960s.

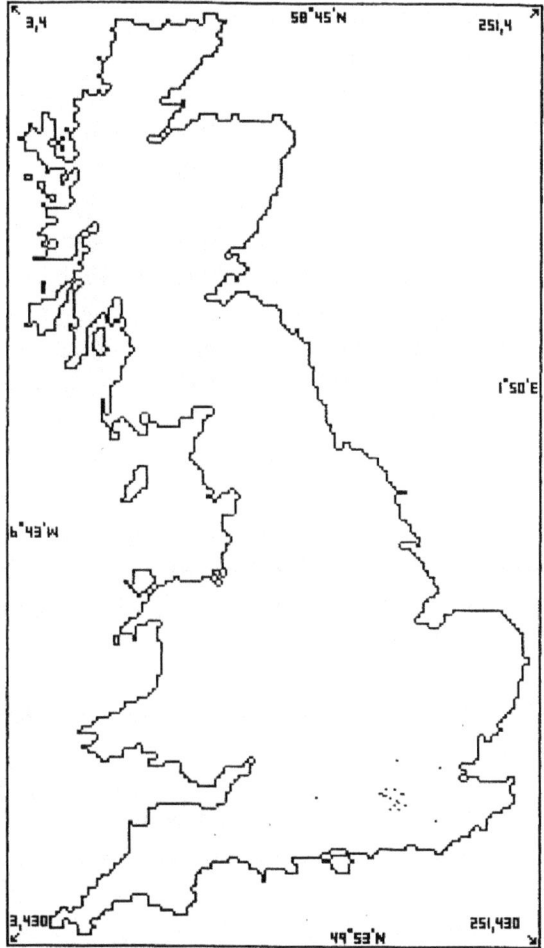

Fig. 7. Reports of ABCs from mainland Britain during the 1970s.

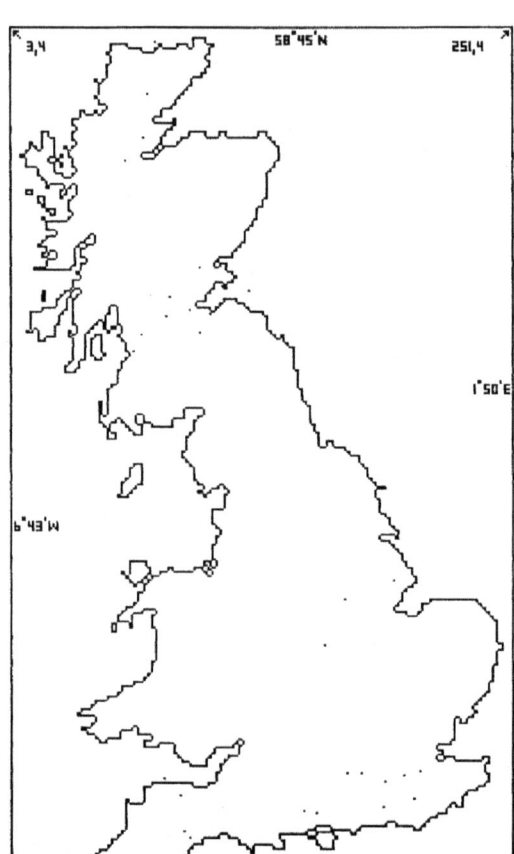

Fig. 8. Reports of ABCs from mainland Britain during the 1980s.

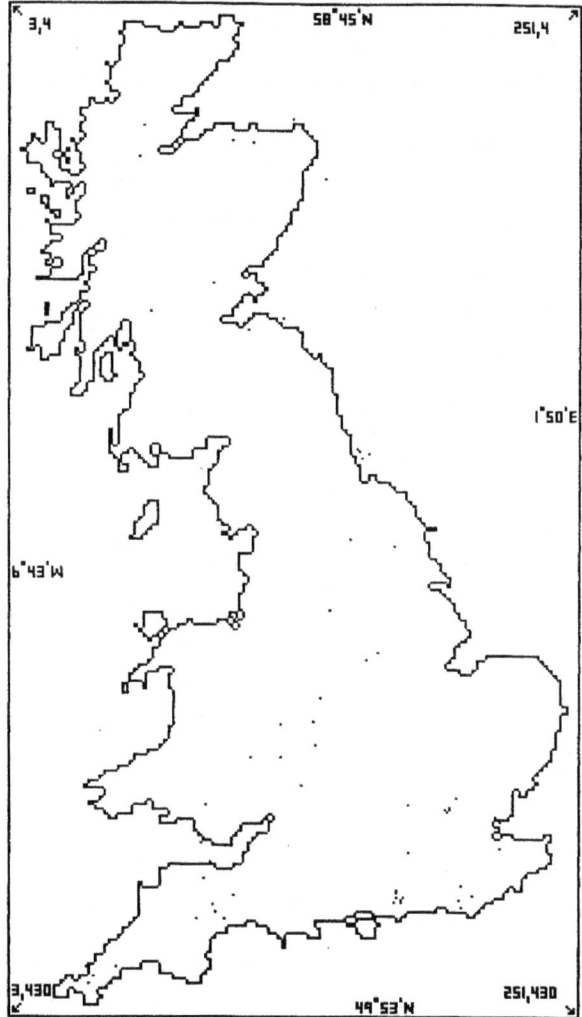

Fig. 9. Reports of ABCs from mainland Britain during the 1990s.

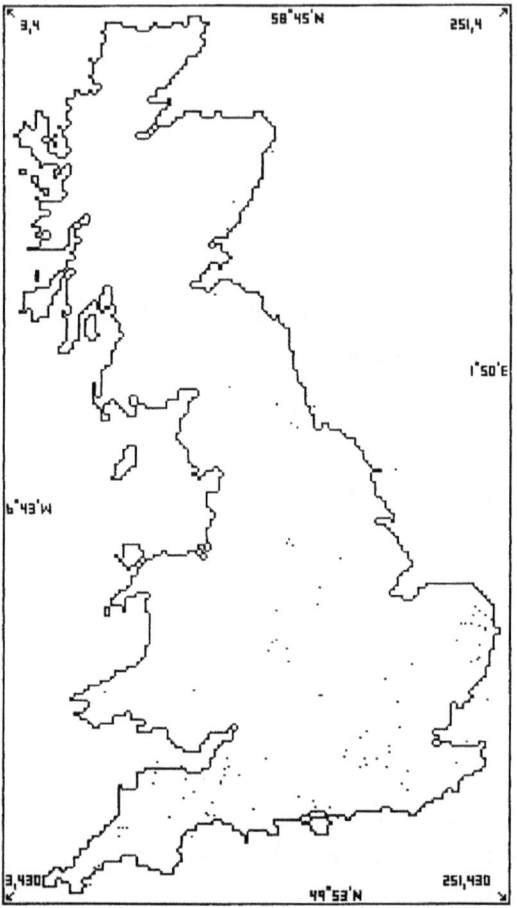

In the 1990s the reports again have a wide national distribution with a relatively high frequency of reports from central and southern England. There are concentration of reports from the Pennines, the south-west, the Home Counties, and for the first time from Norfolk. The 'Peak Panther' and the 'Beast of Bodmin' are mystery felids currently attracting media attention.

The animation demonstrates a number of trends. The first is a steady increase in the number of reports over the years. There appear to always be reports from the south-east of England. Reports from Scotland tend to have occurred during the 1970s and 1980s, with relatively few in this decade so far. Reports from Norfolk appear to be absent prior to 1990. Generally there is a concentration of reports around London during the 1960s. During the next two decades the reports increase in number and expand outwards away from the capital. A further increase in the number of reports during the first half of the 1990s is coupled with a slight contraction in the locations towards the south-east again.

Discussion

It must be stressed that to date this entire project has been conducted totally independently with no collaboration with other researchers in this field. It is almost certain that the reports investigated here represent only a fraction of those on record and there has been no determination of any bias in the sample studied. The conclusions presented below are (unless otherwise stated) based on the assumption that the sample of reports is a fair and reasonable representation of those as a whole, and that the reports are accurate.

1. Twentieth century reports from mainland Britain

Over the century as a whole the number of reports are widespread suggesting that the phenomenon of ABCs is a national one. The land use of the seven regions identified is varied[25], but may be grouped into two major environmental types. 'Wilderness' areas (the Scottish highlands, the Pennines and Devon) and regions of high population density (Edinburgh/Glasgow, Tyneside and London). The exception to this is Norfolk, which is discussed later.

The 'wilderness' environments are relatively remote places with little or no human habitation. Many species of felid could survive reasonably undisturbed in such areas. Aside from these 'wildernesses' the British Isles is heavily overpopulated with a large number of urban areas only a short distance from open countryside. Two major features of urban areas - extensive 24 hour lighting and a high density of pairs of eyes - greatly increase the likelihood of wild animals being observed here.

2. An animation of reports from the 1960s, 1970s, 1980s and the first half of the 1990s.

The increase in the number of reports over the decades may reflect several different trends. Aside from the possibility of sampling bias, reports in *Fortean Times*[27,28] suggest ABC sightings for both 1993 and 1994 were up on those for the previous years. An increase in public interest and awareness may also be contributing to the rise in the number of sightings as people's reluctance to report strange experiences wanes.

Another explanation for the apparent increase in the number of sightings is that there are more cats to be seen. Alternatively the number of cats has changed very little, but their behaviour has become less elusive. They may no longer be as shy of humans as they once were.

The results for the 1960s suggest that the British ABC phenomenon was at that time restricted solely to the south-east of England. The 'Surrey Puma' and 'Shooter's Hill Cheetah'[2] would account for many of these sightings. These two mystery cat episodes brought the whole concept of unknown felids roaming the British countryside to the public's attention and it is likely that the trends shown by the 1970s and 1980s models reflect to a certain extent this increase in public interest and awareness.

Many of the 1970s reports are from densely populated areas, but there is also an increase in reports from the Scottish Highlands and the Devonshire moorlands - two areas where large felids could probably survive quite comfortably.

The 1980s model is most marked by the sudden doubling in the number of reports. In 1976 the Dangerous Wild Animals Act was passed which placed severe restrictions upon the keeping of certain species, including large cats. It is possible that many of the animals kept privately prior to the act found their way into the wild following its introduction. It may have been assumed that the cats that had escaped would very quickly die. However the survival for nine months of a clouded leopard (*Neofelis nebulosa*) that escaped from John Aspinall's Howletts Zoo in 1975 before eventually being shot[1,2] gives no reason to suppose other escaped felids would fare any worse.

A further increase in the number of reports is observed in the first half of the 1990s. One interesting observation is (apparently) the first reports from Norfolk. This may reflect new cats being escaping into this area, or alternatively the migration of animals from other areas.

Problems with the system
Catmap! currently displays a number of minor problems outlined below:

1. Sampling bias - to be overcome by discussion with other researchers
2. Map accuracy - the map was drawn by hand and subsequently enlarged. Accuracy could be improved by scanning an image into the computer.
3. Accuracy of the plotting algorithms - map coordinates are calculated by hand from an atlas and transformed into pixel coordinates by the computer. They are then rounded to the nearest integer for plotting. This, combined with inaccuracies in the map, causes certain locations to be recorded in the sea. The overall accuracy however has been tested and I am satisfied with the performance of the system.

These problems do not interfere greatly with the performance of *Catmap!* and will be addressed in later versions.

Overall

There are many possible explanations for the ABC reports described here. Some may be deliberate hoaxes. Many cases may be misidentification of dogs or domestic cats. Witnesses may be quite innocently seeing what they want to believe rather than what is actually there. The witnesses may of course be seeing genuine cats. If this is so, the trends reported here may be interpreted as:

1. An increase in the number of cats - through escape or breeding
2. A movement of these animals around the British countryside
3. A possible tendency for the animals to become less wary of humans
4. An increase in public interest and awareness.

Future work

It is hoped to increase the size of the database in the catalogue 'UK Felid Historical' in order to perform a more comprehensive study. Aspects of annual migration, seasonal migration and coat colour distribution will be modelled.

References

1. McEwan, G. J. 1986. *Mystery Animals of Britain and Ireland*. Robert Hale, London.
2. Shuker, K. P. N. 1989. *Mystery Cats of the World: From Blue Tigers to Exmoor Beasts*. Robert Hale: London.
3. Francis, D. 1993. *The Beast of Exmoor: and other mystery predators of Britain*. Jonathan Cape, London.
4. Healy, T. and P. Cropper. 1994. *Out of the Shadows: Mystery Animals of Australia*. Ironbark: Chippendale.
5. Kitchener, A. 1991. *The Natural History of the Wild Cats*. Christopher Helm: London.
6. Alderton, D. 1993. *Wildcats of the World*. Blandford: London.
7. Shuker, K. P. N. 1993. *The Lost Ark: New & Rediscovered Animals of the 20th Century*. HarperCollins*Publishers*: London.
8. Shuker, K. P. N. 1993. The Lovecats. *Fortean Times* **6 8**, 50-51.
9. Dash, M. and B. Rickard. 1989. Pussies Galore. *Fortean Times* **5 3**, 8-9.
10. Chen, A. 1990. I was mauled by the Beast of Exmoor. Letter to *Fortean Times* **5 5**:69
11. Maloret, N. 1990. Swamp Cat Fever. *Fortean Times* **5 5**, 44-46
12. Anon. 1992. Mystery Cats Again. *Fortean Times* **6 3**, 20
13. Dash, M. 1992. Mystery Moggies. *Fortean Times* **6 4**, 44-45.
14. Sunday Times 7/11/93
15. Anon. 1994a. Wychwood bear roams the woods. *Fortean Times* **7 4**, 15
16. BBC Radio 2 News 16:00 20/01/94
17. Halstead, R. and P. Sieveking. 1994. An ABC of British ABCs. *Fortean Times* **7 3**, 41-44.
18. Magurtzey, Y. 1994. In the shadow of the panther. Letter to *Fortean Times* **7 3**, 63.
19. The Times, Saturday 12th March 1994

20. Williams, J. 1994a. Newsfile: Mystery Cats. *Animals & Men* **2**, 4-5.
21. Williams, J. 1994b. Newsfile: Mystery Cats. *Animals & Men* **3**, 8-10.
22. Williams, J. 1994c. Mystery Cats - where do we go from here? *Animals & Men* **3**, 20-25
23. Williams, J. 1994d. Newsfile: Mystery Cats. *Animals & Men* **4**, 5-6.
24. Fullard, H. (ed) 1976. *Philips' Modern School Atlas* (74th edn). George Philip & Son Ltd., London.
25. *Atlas of the World*. John Bartholomew & Son Ltd., Edinburgh (1986).
26. Ordnance Survey. 1994. *Motoring Atlas*. Hamlyn, London.
27. Anon. 1994b. The *Other* FT Index. *Fortean Times* **73**.
28. Anon. 1995. The *Other* FT Index. *Fortean Times* **79**.

BLACK SHUCK OF NORFOLK

Hetz, Hetz! let slip this famous coursing hound,
The Huntress of the Moon is stalking earth.
Out of the Night and Mist her dark hound leaps
To chase the flying man across his land.

Now collared, kennelled, named, he can be set
To prick for game in beechwood forest thick;
Who flies the questing pair is further lost,
For *Nacht und Nebel* runs before the pack.

The hunted hunter flees the one desired;
That which he follows, follows him again:
But when the maiden can the dogs call in,
Is he the more afraid to know her power?

Harry the fearful as he runs from air,
Frighten the harrier who pursues his fear,
Panic the two who circle round the trees,
Chaste chaser, let them both escape the shot.

So yet the ritual death-in-life still runs,
And if the lie is true, how lies the game?
Can he be hunted for his craft and sleight?
Or his naive and candid joy in sport?

But for the silver archer in the void,
Game is true worth of faith and loyal heart:
This is an earthling, heir to all her realm,
Who will his own true comrades welcome home.

The Mystery Bird from Hiva-Oa

by Michel Raynal.

Cryptozoology is not only interested in Nessie and Bigfoot. At least one hundred and forty other animal forms unknown to science are relevant to cryptozoology [1]. One of the less known cryptids is a wingless bird said to live on Hiva-Oa in the Marquesas Islands.

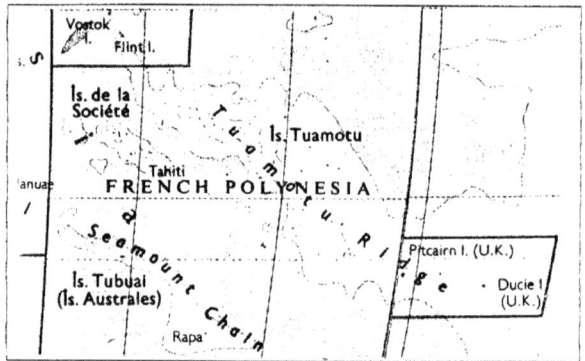

French Polynesia

In 1937, Norwegian explorer Thor Heyerdahl, (the famous hero of the Kon-Tiki expedition across the Pacific on a Balsa Raft), was on Hiva-Oa, riding in a mountain forest, together with a Polynesian native named Terai. In his book Fatu-Hiva, back to nature (1974), he related a strange observation:

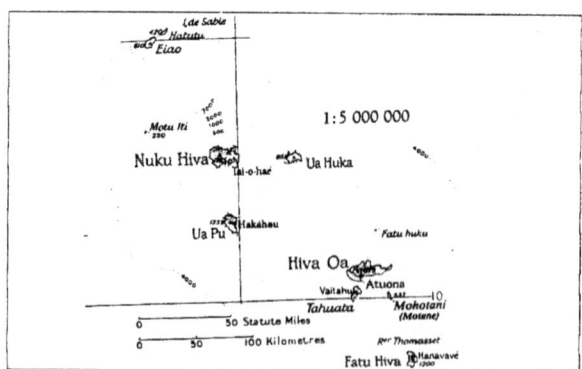

The Marquesas Islands

"Suddenly Terai halted his stallion and pointed to the trail in front of him. We had reached a hillock with low ferns, and a bird without wings was watching us. Then it ran faster than a little hen along the trail and disappeared light lightning into a sort of tunnel between the dense ferns. We had heard of this wingless bird, a strange species quite unknown to Ornithologists. The islanders had often seen it but had never managed to capture one, as it always disappeared into holes and tunnels at great speed" [2].

In 1980, I was able to contact Thor Heyerdahl; 43 years after this sighting. He did not remember the appearance of the bird but he wrote, nevertheless:

"My memory of the wingless bird of Hiva-Oa has left me with the impression that it was considerably bigger than a sparrow and rather the size of a long-legged sea gull" [3].

In 1956, French explorer Francis Maziere also heard of this bird on Hiva-Oa, as he writes in his book 'Mysterieux Archipel du Tiki' (1957):

"In Hiva-Oa Island, I heard from an old Norwegian sailor, and from some natives of a curious animal, named Koau, which looked like a kind of Kiwi.

According to the old Lee (sic) who had made after one on horseback but failed to capture it, as it was running too fast, the animal was as big as a cock, its fur (sic) was purplish-blue, its beak yellow, as well as its long and strong legs. It only had stumps of wings.

This description sounded so extravagant that I did not pay attention to it, until I came across a New Zealand magazine, relating, with photographs in support - which allowed us to show them to the natives - the discovery, by a mountain expedition of an unknown group of noctunis (sic), which had taken refuge on the edge of a glacier. The animal was the same. What should be emphasised is that, being unable to fly, this species could have been transported only by migrations in canoes. Unless of course the famous continent of Man (sic) did really exist in the Pacific area". [4].

Several errors should be noticed in Maziere's text, which were certainly made by the printer rather than by Maziere himself.

The "old Lee" is Henry Lie, a Norwegian sailor living in Hiva-Oa, and a friend of Thor Heyerdahl's. Then, "Fur" is a curious integument for a bird. New Zealand kiwis have feathers which look like hairs. In fact "fur" is a mis-spelling of 'plumage'. Both words in French are similar enough ('pelage' and 'plumage').

EDITORS NOTE: In Chapter Ten ('*The Moa-a fossil that may still thrive*') of '*On the Track of Unknown Animals*' (1995 reprinting) Bernard Heuvelmans notes on several occasions that both Kiwi's and the (presumably) extinct Moas have remarkably hairy, or even furry, plumage and he also makes some interesting comments on the word 'Moa' in the traditional Maori language:

"It is true that the word 'Moa' is sometimes used for a hairy, waving, flaccid and coarse growing seaside grass (Spinifex hirsutus), but it may well have been named

after the Moa's hairy plumage. And not only its leaves, but also a local variety of flax (Phormium), which provides a textile fibre, are called raumoa or "moa's feather"..."

"Noctunis" is a misspelling for "Notornis", as remarked by Gabriel Linge [5] (1972), and finally, the so-called continent of Mu (invented by James Churchward), should not be confused with the Isle of Man, between Great-Britain and Ireland.

In 1977, I spoke of the previous reports to Bernard Hewuvelmans, the father of Cryptozoology. He forwarded them to ornithologist Jean-Jacques Barloy, who mentioned them in his book *Merveilles et Mysteries du Monde Animal* (1979). He added some details on the mystery bird that were given to him by Francis Maziere himself:

"The species is now extinct, the victim of a foolish overhunting. Frenchmen knew the bird and hunted it. Bones are said to be in some tombs" [6].

The name of Koau, reported by Maziere, means 'peak' in Marquisian, according to Dordillon's dictionary (1904), but the similar sounding name of Koao is translated as 'a kind of bird' [7].

More information on the Koao is given by doctor, Louis Rollin in his book 'Les Isles Marquises' (1927):

"KOAO: burrowing bird which lives in the mud of the plantations of "ta'o" (taro). At the slightest noise it burrows a hole, which makes its capture difficult". [8]

Is It The Spotless Crake?

Porxana tabuensis

In 1979, Jean-Jacques Barloy [6], suggested that this bird might be the Spotless Crake (*Porzana tabuensis*). It is a rail living on many of the islands of Polynesia, including the Marquesas Islands. It is 15 to 20 cm (6-8 inches) long, black, with a rudimentary tail, stumpy wings, red eyes and feet, and a light brown bill. It runs very fast and would rather run from danger than fly. In the Marquesas, it would survive only in some valleys of Ua-Pou and Fatu-Hiva.

Jean-Claude Thibault, whose ornithological expedition to the Marquesas Islands failed to observe this species, wrote:

"Several Marquesians assured us that the bird, when seen, burrows a hole in the mud and dives into the taraudieres (tapped out passages)" [9]

It is exactly, and almost in the same words, what doctor Rollin wrote [8], about the Koao in 1927.

Even the very native name of the bird supports Barloy's hypothesis: According to Holyoak and Thibault [10], Koao is the Marquesian name for the spotless crake. However, Barloy (1979) had some reservations about his own hypothesis that this small rail might be the mystery bird sighted by Heyerdahl and Lie:

"Is that species the Koau? Its ecology and its behaviour are similar, but its size is much smaller. The mystery remains" [5].

Let us remember that Heyerdahl's bird was *'the size of a long-legged sea-gull'*, and Lie's one *'as big as a cock'*. Apart from the size, the long legged feature confirms Henry Lie's statement that the bird that he saw had *'long and strong legs'*, quite unlike that of the spotless crake. In addition, the black colour of the spotless crake differs greatly from the purplish-blue hue of the unknown bird.

If Koao is really the name of the spotless crake, the confusion may have been introduced by Maziere: having recorded Lie's observation of a wingless bird, he may have asked the natives if they knew such a bird, and they spoke to him of the spotless crake, alias Koao. Unless, of course, the name Koao is given to both of these flightless birds by the natives.

Is it the Takahe, the Moho, or an unknown endemic rail?

In an article for the Bulletin de la Societe d'Etudes des Sciences Naturelles de Beziers (1981) ([11]), I concluded that the mysterious bird of Hiva-Oa was very similar to the Takahe from South Island, New Zealand.

About 150 years ago, Maori natives of the North Island, of New Zealand reported the recent existence of a flightless bird, which they called Moho, distinct from Moas and Kiwis. They had, however hunted it to the brink of extinction.

In 1847, Walter Mantell obtained some subfossil bones of Moho from Waingongoro (North

Takahe *(Notornis (Porphyrio?) mantelli)*

Island), which he forwarded to anatomist Richard Owen, in London, precising that the bird was also called Takahe in South Island. It was a large rail with rudimentary wings, which Owen described as *Notornis mantelli*.

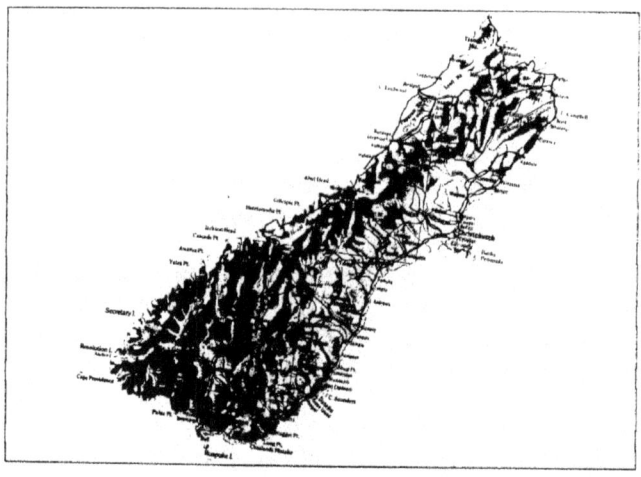

The South Island of New Zealand

In 1849, the dogs of seal hunters caught a big flightless bird on Resolution Island, off South Island: Its plumage was purplish-blue, somewhat green on the back and the wings, it had a thick, red bill and strong, red legs. This bird belonged to the same species as the one examined by Owen. Supposed to be extinct from North island before it was discovered alive the Notornis survived in South Island!

In 1850, a Maori caught a second Takahe on Secretary Island in the same area. In 1879 the dog belonging to a rabbit hunter caught another specimen near lake Te-Anau. It appeared that they were two different sub-species. The Moho of North Island, *(Notornis mantelli mantelli)*, only known in a sub-fossil state, and the Takahe of South Island *(Notornis mantelli hochstetteri)*.

A fourth specimen was also caught by dogs in 1898, near lake Te-Anau. the Takahe was then considered extinct, but in 1948, Geoffrey Orbell rediscovered this bird near lake Te-Anau, and it is now protected by New Zealand wildlife services.

In 1981, when I published my first article on this problem [11], the 'identikit' picture of the mystery bird from Hiva-Oa made me think of the Takahe.

The data was similar for both birds.

Ecology:
(A similar habitat of forests and high mountain grasslands).

Ethology:
(They both run and hide at the slightest provocation).

Size:
(Both birds are the size of a cock or a sea-gull or a cock).

The description also corresponds closely. Both birds have rudimentary wings, strong legs, purplish blue plumage and similarly coloured bills and legs.

I also noted the rapid running gait of the Takahe, which allowed it to remain incognito until as recently as 1948. Only four specimens were known previously to this year, and three of these were caught by dogs, animals able to follow the birds through their natural habitat. This would account for the similar elusiveness of the mystery bird of Hiva-Oa.

Even the good tasting flesh is a feature shared by the two birds. It was the cause of the extinction of the Moho (the North island Notornis), and of the rarity, (if not the recent extinction of the mystery bird).

Last, but not least, Maziere said that his Marquesian informants had recognised their mystery birds on photographs of 'noctunis' (read Notornis), - irrespective of whether or not its native name is really Koao.

There was just one little difference. The legs and beak of the unknown bird were said to be yellow, whereas they are red in the Takahe. these colours are, however, very close in the visible spectrum, particularly if one remembers the conditions in which they were observed, in the darkness of a forest, with the bird running at a considerable speed, making it impossible to examine a specimen at leisure.

I thus proposed [11] that the Marquesian bird was the Takahe (or the Moho), brought from New Zealand by Maori natives in their dug-out canoes. Incidentally Gabriel Linge made the same hypothesis in his book 'Nouvelle-Zelande, Terre des maoris' (1972), after he quoted Maziere and gave the correct spelling of the so called 'noctunis':

"An unexpected argument which could support the hypothesis of early contacts between New Zealand and the islands of Eastern Polynesia is the following one. A bird believed to be extinct for a long time, the notornis, has been found again - and is living - in only two places: Marquesas Islands and New Zealand" [5].

In 1981, my preferred explanation was an 'importation' from New-Zealand, because, as explained by Maziere, the bird could not reach Hiva-Oa by flying, (it is flightless), nor by walking (as there never was any land of Mu!) - nor by swimming!

I neglected the possibility, however, that the rail once reached Hiva-Oa by flight, when it was still able to fly, and that it later evolved into an apterous form, as is the case with most island rails. It is a well known evolutionary process, which one explains by the absence of predators in these islands ... until man arrived. I only alluded to this possibility at the end of my 1981 article, when I concluded:

"One can also imagine that the Koau is another species of porzana, or even a rail of a still unknown genus". [11].

Jean-Jacques Barloy again alluded to the unknown bird from Hiva-Oa in his 1984 book *'Les survivants de l'ombre'*, and he was less sure that it was the spotless crake:

"M. Raynal prefers to refer to the Koau as another rail, the Notornis or Takahe from New Zealand (.......) why not?" [12].

In 1986 Bernard Heuvelmans listed 'my' bird in his famous check-list of the one hundred and forty forms still unknown to science, relevant to cryptozoology, and he mentioned my hypothesis, but he also expressed his own, less restrictive opinion:

"It has been suggested that it is closely related to the New Zealand Takahe, and thus a species of Notornis (Raynal 1980-81). What can be put forward more safely is that it looks, indeed like a rail, but is larger than the local Porzana tabuensis, thought to be locally extinct and surviving only on other islands in the same archipelago". [1]

Meanwhile, according to Olson's work (1973), the genus Notornis had become a synonym with Porphyrio [13]. It is true that the purple swamphen (Porphyrio porphyrio) looks like the Takahe in its shape, size (only somewhat smaller), colour (purplish-blue), behaviour etc., and

The Purple Swamphen or Gallinule
(Porphyrio porphyrio)
Picture couyrtesy Dr. Shuker

all which has been said for the Takahe can be said for the purple swamphen as well.

Moreover. the same native name of Moho, given by the Maoris of North Island to the extinct cousin of the Takahe *(Porphyrio mantelli hichstetteri)*, is often given to the spotless crake *(Porzana tabuensis)* in many islands of the tropical Pacific [10]: Moho in Tuamotu, Meho in Tahiti, Mo'o in Atiu, etc. This demonstrates a link between both of these flightless birds in Polynesian culture; so the same word Koao might be applied to the spotless crake in the Marquesas Islands where it is still living (in Fatu-Hiva for instance), and to the unknown bird related to the Moho in Hiva-Oa. Koao would thus mean something like 'flightless bird' in Marquesian.

Recent Discoveries in Hiva-Oa.

In 1990, English Zoologist Dr Karl P.N.Shuker summarised the affair in an article on the birds still unknown to science. He mentioned my first hypothesis, recalled the sensation created by the rediscovery of the Takahe in 1948, and concluded:

"Who knows, perhaps some future ornithological investigation on Hiva-Oa may engender a repetition of history". [14].

This had, in fact, almost happened two years earlier, after the discovery by David W. Steadman, of the New York State Museum, of subfossil bones about a thousand years old, of a new species of rail, named Porphyrio paepae, in archaeological sites of two islands in the Marquesian archipelago, Tahuata and Hiva-Oa!

They have been discovered in pae-pae (hence the name given to the species), platforms used as a base for various dwellings, often kitchen middens and sometimes religous sepultures. This confirms Maziere's information reported by Barloy:

"The species is now extinct, the victim of a foolish over-hunting. Frenchmen knew the bird and hunted it. Bones are said to be in some tombs". (9).

Though the external appearance of this rail cannot be determined from its bones, it is clear that *Porphyrio paepae*, which obviously looks like the Takahe *(P.mantelli)*, and the mystery bird are one and the same species. It is likely to have been purplish-blue in colour and to have had a yellow bill and feet. My cryptozoological analysis of this file in 1981, despite its errors, was not so far from the truth.

The discovery of *Porphyrio paepae* represents an eastward range extension of the genus Porphyrio of three thousand, two hundred kilometers. The purple swamphen *(P. porphyrio)* only reaches the islands of Western Polynesia, whereas a species extinct at the end of the 19th Century *(P.albus)* lived on Lord Howe island.

Reconstruction of P.paepae by the author

Steadman (15), states that the genus Porphyrio has never been recorded either living, or fossil, in Eastern Polynesia. In fact, Maziere,[4] in 1957, Linge[5], in 1972, Raynal[11], in 1980-1, Barloy[12], in 1984 and Heuvelmans (1), in 1986 had already alluded to a link between the mystery bird from Hiva-Oa and the Takahe from New-Zealand (Porphyrio mantelli), but I can understand how he was not aware of these rather obscure references. On the other hand, ornithologist Lionel W.Wiglesworth, as early as 1890, mentioned the presence of *'Porphyrio spp.'* in Raiatea (Iles Sous-le-Vent, Eastern Polynesia), from observations not confirmed since.[16]

Steadman has suggested that fossils of the genus Porphyrio will be discovered elsewhere in Eastern Polynesia[15]. Apart from Raiatea, I propose to search in Tahiti, where James Morrison, boatswain's mate aboard the Bounty, mentioned two centuries ago, *'a large bird, nearly the size of a goose, which is good food'*, never observed near the sea or in the low lands of Tahiti. (This story is, of course, less well known than the famous mutiny!). Derscheid (1939) thought that it was a true goose[17], but the description is too vague to be so accurate. What can be said, however, is that no such large bird is presently known from Tahiti.

The confirmed existence of the mystery bird from Hiva Oa and the checking of its presumed zoological affinities provide new evidence of the efficiency of cryptozoological research following the methodology defined by Heuvelmans[18] in 1988. This I emphasised in an article written with Michel Dethier for the Bulletin Mensuel de la Societe Linneenne de Lyon in 1990.

Nowadays, if any field cryptozoologists go to Hiva-Oa, they should show the Marquesians a reconstruction of P.paepae. This might allow us to collect new, and more recent sighting reports, and also to accelerate the discovery of a living specimen, if by chance it still survives. This method was carried out successfully by Indian ornithologist Salim Ali for the Jerdon's courser (Cursorious bitorquatus), which was supposed to have been extinct for a century and a half before its rediscovery in India in 1986.

Hopefully, may it be the same for the (now less) mysterious bird from Hiva-Oa!

REFERENCES.

1. HEUVELMANS Bernard, *'Annonated Checklist of Apparently Unknown Animals with which Cryptozoology is concerned'*. Cryptozoology, 5 (1986). pp 1-26.

2. HEYERDAHL, Thor, *'Fatu-Hiva, back to nature'*. (London, George Allen and Unwin, 1974 p.225).

3. HEYERDAHL, Thor, private communication (letter of 17 December 1980).

4. MAZIERE, Francis, *'Mysterieux Archipel du Tiki*. (Paris, Robert Laffont, 1957, p.261).

5. LINGE, Gabriel, *'Nouvelle-Zelande, Terre des Maoris'*. (Paris, Robert Faffont, 1972 p.62).

6. BARLOY, Jean-Jacques, *'Merveilles et Mysteres du Monde Animal'*. (Geneve, Famot, 1979, pp. 115-117).

7. DORDILLON, J.R., *'Grammaire et Dictionnaire de la Langue des Marquises'*. (Paris, Institut d'Ethnologie de l'Universitie de Paris, 1904, pp.225).

8. ROLLIN, Louis, *'Les Isles Marquises - Moeurs et Cotumes des Anciens Maoris des iles Marquises'*. (Paris, Societe d'Editions Geographiques, Maritimes et Coloniales, 1927, pp.50-1).

9. THIBAULT, Jean-Claude, *'Notes ornithologiques polynesiennes. II - Les Marquises'*. Alaudau, 41 : 3 (1973), pp. 301-316.

10. HOLYOAK, D.T., et THIBAULT, J.C., *'Contribution a l'etude des oiseaux de Polynesie orientale'*. Memoires du Mueum d'Histoire Naturelle, Serie A., Zoologie, 127 (1984). pp. 1-196.

11. RAYNAL, Michel, *'Koau, l'oiseau insaisissanble des iles Marquises'*, Bulletin de la Societe d'Etude des Sciences Naturelles de Beziers, 8°: 49 (1980-81), pp. 20-6.

12. BARLOY, Jean-Jacques, *'Les survivants de l'ombre'*, (Paris, Arthaud, 1984, pp. 197-99).

13. OLSEN Starrs L., *'A classification of the Rallidae'*, Wilson Bulletin, 85, (1973), pp. 381-416.

14. SHUKER, Dr Karl P.N., *'A selection of Mystery Birds'*, Avicultural Magazine, 96:1 (1990), pp 30-40.

15. STEADMAN, David W., *'A new species of Porphyrio, (Aves:Rallidae) from Archaeological sites in the Marquesas Islands'*. Proceedings of the Biological society of Washington, 101°; 1 (1988), pp. 162-170.

16. WIGLESWORTH, Lionel W., *'Aves Polynesiae'*, Abhandlungen und Berichte des Koniglichen Zoologischen und Anthropologische und Etnographischen Museums zu Dresden', 3 : 6 (1980-91), pp. 1-92.

17. DERSCHEID, M., 'An unknown species - The Tahitian Goose (?)', Ibis (1939), pp. 756-760.

18. HEUVELMANS, Bernard, 'The Sources and Method of cryptozoological research'. Cryptozoology, 7 (1988), pp. 1-21.

19. RAYNAL, Michel et DERTHIER, Michel, 'Lezards geants des Maoris et oiseau enigmatique dea Marquisiens: la verite derriere la legende", Bulletin Mensuel de la Societe Lineenne de Lyon, 59°: 3 (Mars 1990), pp. 85-91.

EDITOR'S NOTE: This article first appeared in English in issue 5 of The Crypto Chronicle, but with only one illustration and no references.
It is reprinted with kind permission of M. Raynal.

* * * * *

Big Cats in Norfolk during 1995.

by Justin Boote (Norfolk Correspondent).

On Tuesday 2nd May 1995 a large felid was allegedly seen at Watton Airfield, approximately fifteen miles from Norfolk. It was reported to me by Mrs B.Muskett who described it as 'bigger than a labrador' (which is what she originally thought that it was), and a beige colour with a very long tail which curled upwards at the tip. There were no distinguishing features. She said that it resembled a 'mountain lion'. The animal was walking by some tennis courts and upon seeing Mrs Muskett promptly loped off.

There were some children playing nearby and they were all aware of the animal.

On Thursday 4th May it was back and in the same location. A young girl saw it in the morning and Mrs Muskett saw it again later that day. Mrs Muskett was walking her dog and was aware of the cat moving towards her. She started to run, and the cat began to chase her in 'a 'playful manner'. Again there were children playing nearby, but it kept away from them. She looked for any evidence of the animal the next day but the grass had been cut, and there was none.

On Sunday 7th May it reappeared again, and again Mrs Muskett was witness to it. She was exercising her dogs in the same area and the cat was basking in the warm weather, simply lying on the ground asleep. She walked past quickly, and it carried on watching her before disappearing later. She reported it to the police but it was gone before they could get there. The cat was seen by other people the following week.

One person to see the animal was a farmer (name unknown) at Wick Farm which was where Raymond Trew took the 1994 photographs.

Another black feline was seen in Ashill just a couple of miles away in June 1994. No escaped felids have been reported although there is scattered forestry nearby and the cat may have been searching for food away from its normal hunting grounds.

*****

When the Boat Comes In...

by Stuart Leadbetter.

When I was a small boy during the 1970's, I remember watching a BBC drama series which was very popular at the time about life in a fishing port in North-East England. The programme was of more interest to my parents than it was to me. Being born and brought up in my home town of Fleetwood, they would have been familiar with the everyday activities of a fishing port. It is only within the past few years that I have become interested in the history of Fleetwood, and by then its heyday as a fishing port had long since passed.

The site of my research into Fleetwood's past has been the local library, and the source of the information I have been able to uncover has been the back issues of two local papers - The Fleetwood Chronicle and the long defunct Fleetwood Express - kept there on microfilm. As you might have guessed, I was not looking for newspaper reports on the normal, everyday activities which happen in a town or city, but instead I was looking for things of a cryptozoological nature. Most of what I found was concerned with unusual species of fish brought in by the fishing fleet.

The first article that I discovered came from the 17.7.1897 issue of the Fleetwood Express. A small paragraph told the reader that a sturgeon weighing 137 pounds (62 kgs), and about seven feet (2.1 m), was caught by a small prawner named 'The Charlotte-Anne'. The fish had been caught on the previous wednesday, the 14th July. When dragged on board the sturgeon still had a lot of life in it, and struck one of the crew-members on the wrist with its tail. When finally dispatched and brought back to land, the fish was sent to London, presumably after having been bought by person or persons unknown.

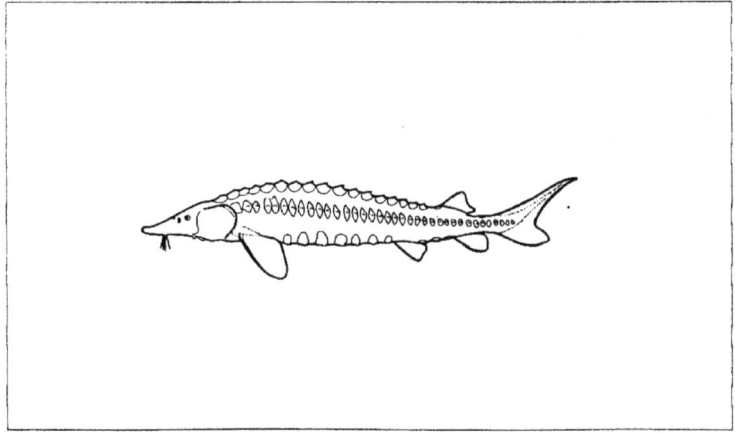

Sturgeon (*Acipenser sturgo*)

EDITORS NOTE: Sturgeons, together with all whales, porpoises, and dolphins caught in English waters, whether dead or alive, are deemed Royal Fish, and are the property of the Crown, except where the carcass has been washed ashore or stranded within the limits of a Manor, in which the title of Royal Fish has passed from the Crown to the Lord of the Manor. The law is slightly different in Scotland, where it does not apply to whales under 25 feet in length or of the species *Hyperodon ampullatus (Forster)*.

The liability for disposal or burial of carcasses belonging to the Crown rests with the DTI. Sturgeons however are edible, and are considered a delicacy, and whereas in the 20th Century the Crown right to specimens of *A.sturgo* is often waived, a hundred years ago, it is likely that the Fleetwood fish would have been sent to London for Queen Victoria, or at least her household.

In this brief paragraph no details were given as to where the sturgeon was caught, but it could only have been *A.sturgo*.

From the length of this fish it is possible to estimate its age when caught. In their book 'Freshwater Fishes', P.S Maitland and R.N.Campbell present a growth curve of a sturgeon in the St. Lawrence river of North America. The two axes of the graph are 'Age in years', on the horizontal axis, and 'Length in mm's' on the vertical axis. By locating the length of the sturgeon caught by the Charlotte Anne on the vertical axis, and then reading across and then down to the horizontal axis, the age given is approximately thirty years. (This may not be strictly true. Some authorities consider the sturgeons native to North America to be a separate species (Acipenser oxyrhynchus), and of course, if they are, they may well have a different growth rate to the common sturgeon.

The next fish to turn up in the nets of a trawler was a strange one thought by the members of the ships' crew to be some kind of shark. This occurred on the 4th of May 1905 and was reported in the following saturday's Fleetwood Express.

The length of this peculiar fish was stated to be a yard (0.9 m), and when opened up, fifteen young fish, presumably of the same species were inside. They appeared to be fairly well developed.

Trying to determine what species of shark this fish was (if shark it was) is extremely difficult. I can only find details of five species of shark which are found in British waters which grow to a metre in length and reproduce ovo-viviparously. (This is when the eggs holding the young hatch within the mother and feed independently of her. It is possible that the shark in question reproduced viviparously like the mammals do, but, so far as I can tell, within British waters there are no small shark species that breed in this manner. I may, however, be wrong).

The five shark species which fulfill all five criteria are:

The Angel Shark or Monkfish (Squatina Squatina)

A small shark from the temperate sea beds of Europe, from the coast to depths of 150 metres.

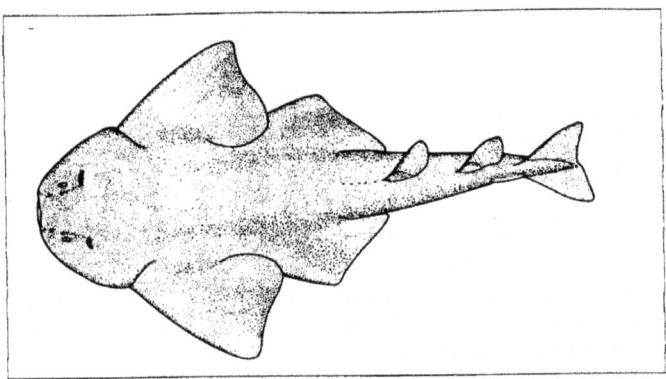

Prefers muddy or sandy bottoms, where it lies with only its eyes showing above the mud. Nocturnal activity. Sleeps on the sea bed during the day. Penetrates farther north in summer. It feeds on bony fish, skates, rays and other flatfish, crustaceans and molluscs and reaches a maximum size of 1.83-2.44 metres.(Xavier Maninguet: 'The Jaws of Death' Harper Collins 1992.

The Piked Dogfish *(Squalus acanthias)*

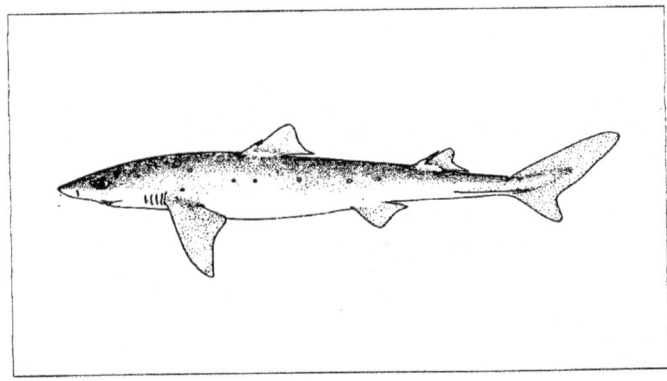

Also known at the Spiny Dogfish or Spurdog this fish is grey with white spots and a white belly, and never exceeds 1.6 m in length. It is a voracious predator which frequents boreal to warm coastal waters across much of the world. It is without doubt the most widespread shark in the world and is fished on an industrial basis. 35,000 were caught during 1978 in Europe alone. Slow and inactive it can gather in enormous shoals. Although it is a coastal creature it cannot tolerate fresh water and during times of heavy rain it moves away from the coasts to areas of greater salinity. Its ideal water temperature is between 7°C and 15°C which explain its seasonal migrations downwards or northwards. One tagged individual was caugfht again seven years later 6,500 km away. (Xavier Maninguet: 'The Jaws of Death' Harper Collins 1992.)

The Sailfin Roughshark *(Oxynotus paradoxus)*

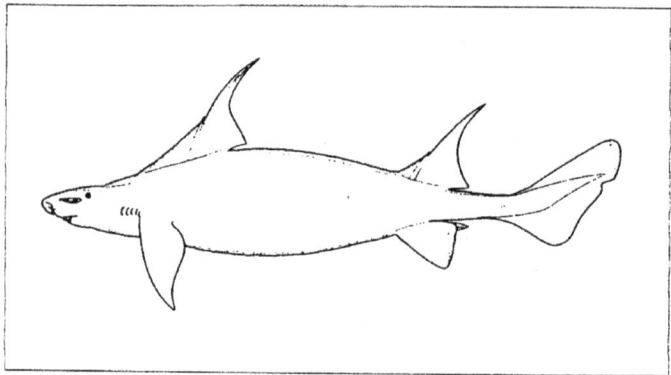

A uniform dark grey or dark chestnut shark which reaches a maximum length of 1.18 m. It has distinctive fins shaped like sails with fins. (Xavier Maninguet: 'The Jaws of Death' Harper Collins 1992.)

The Starry Smoothound (Mustelus asterias)

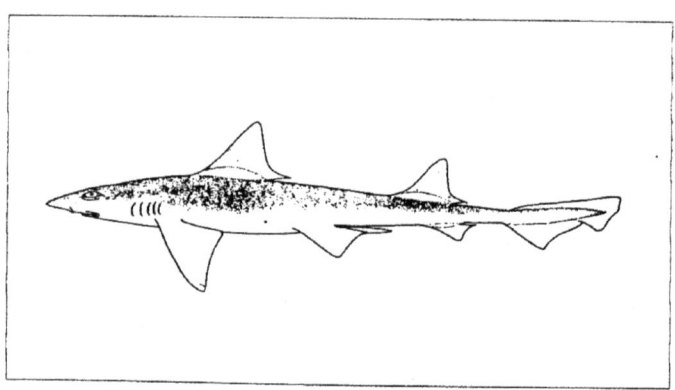

This small, white spotted shark grows up to a length of about 1.4 m and lives off crustaceans, especially hermit crabs. 7-15 pups per litter. (Xavier Maninguet: 'The Jaws of Death' Harper Collins 1992.)

The final contender is the very rare Frilled Shark (*Chlamydoselachus anguineus*).

I am fairly sure that if the identity of the strange fish is not found to be one of the above species, then examination of a more extensive list of sharks would provide the answer. I personally favour the identity of the Piked Dogfish, followed by the Sailfin Roughshark (because it looks so weird) and then possibly even the Frilled Shark.

Unfortunately the crew of the trawler were not even certain that what they had caught was some species of shark. They were trawlermen not Icthyologists and the possibility remains that the fish was of a type unknown to science. Seeing, however, as ninety years have elapsed since the fish was caught, I fear that we shall never know the true identity of the fish dragged up onto the deck of the un-named fishing vessel.

The next two articles I found gave details of the capture of three sturgeons. The first one was landed by the steam trawler 'Desideratus', to the south of the Western Isles, a little less than four weeks after the netting of the strange, shark like fish. It tipped the scales at nearly two hundredweight (101 kgs.), considerably heavier than the one which had been landed in 1897. This large specimen was subsequently bought by Mr Ben Leadbetter, a local fish merchant, who may have been a relative of mine, and was subsequently sold on, probably making Mr Leadbetter a tidy profit. The same man, three years previously had purchased another sturgeon caught by Fleetwood trawlermen, which was subsequently accepted by His Majesty the King. No details about this fish were given in the report.

Yet another sturgeon was captured by the steam trawler 'Elswick' on 21st March 1909, and was duly reported in the following saturday's Fleetwood Express. This fine Royal Sturgeon, weighing 15 stone 8 pounds (98 kgs), was sold for £8.00 to a fish merchant from Manchester and Oldham.

We move forward to 1913 to find a report of a huge Tunny Fish being caught by a Fleetwood Trawler off the south west coast of Ireland. When weighed it tipped the scales at 715 lbs (324 kgs). According to the report published in the 19th September edition of the Fleetwood Chronicle for that year, the Tunny fish is a rare visitor to local waters. It is a member of the Mackerel family and a relative of the Tuna.

Now we come to what I consider the most interesting and rarest fish landed at Fleetwood, we have to go forward to 1961.

In the May 18th edition of the Fleetwood Chronicle a front page story tells of the landing of a Frilled Shark by the trawler 'Red Crest'. when first caught, local trawlermen were baffled as to the fish's identity, but after the local Chief Public Health Inspector had sent it to the British Museum, officials identified it as a Frilled Shark (Chlamydoselachus anguineus). At the time of the fish's discovery the British Museum had no specimens from local waters and they were particularly pleased that the fish was in such good condition, and so large, being six feet (1.8 m) in length and over 14 lbs (6.3 kgs) in weight.

My interest in the fish, fired by this story, I attempted to discover more about it, but I only have one book which supplies me with any further information. *'There are Giants in the Sea'* by Michael Bright, (pp 185-6), describes the Frilled Shark as:

"An ancient shark, almost a living fossil which was found off the coast of Japan and identified for science by Samuel Garman in 1884. It has rarely been seen, but those that have been caught - from deep water off Japan, western Europe, California and South-West Africa - indicate a wide distribution.

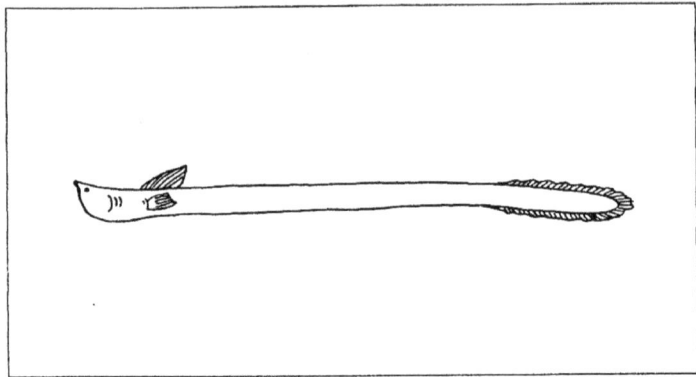

Frilled Shark

The shark is peculiar in being eel shaped. It has a single dorsal fin set far back on its body, a mouth at the front of the head (rather than underneath like most sharks), and six frilly gill slits, the first of which, almost encircles the body".

The most eye opening quote from the author was his guarded assertion that maybe giant frilled sharks swim the seas of the world and are responsible for some reports of sea-serpents:

"Its appearance - the serpentine body and frilled 'mane' - have stimulated several commentators to suggest it as a candidate for the great sea-serpent...that is if GIANT frilled sharks exist in the worlds oceans".....

Over the next two pages Bright quotes eyewitness testimony of sightings of similar creatures to back up his theory.

Four years later another strange fish turned up at Fleetwood and was duly reported soon after in the local paper. It was not perhaps as rare as the Frilled Shark, but is still a very interesting catch. It was a Sunfish (Mola mola). This particular specimen weighed some 172 lbs. (78 kgs.), and was nearly five feet (1.5 m.) in length, although no mention was made in the report to the distance between the dorsal and anal fins.

The trawler responsible for the fish's capture was the 'Robert Hewet', and the fish was taken off the western coast of Iceland. The manager of the local Mac Fisheries, a Mr B.Cheetham bought the fish for the princely sum of £7 10 shillings. Soon afterwards it was put on display at one of the Mac Fisheries shops in Manchester. In finishing the report stated that another Sunfish had been landed at Fleetwood two years previously by the trawler 'Lucida'. This particular fish weighed 150 lbs. (68 kgs.).

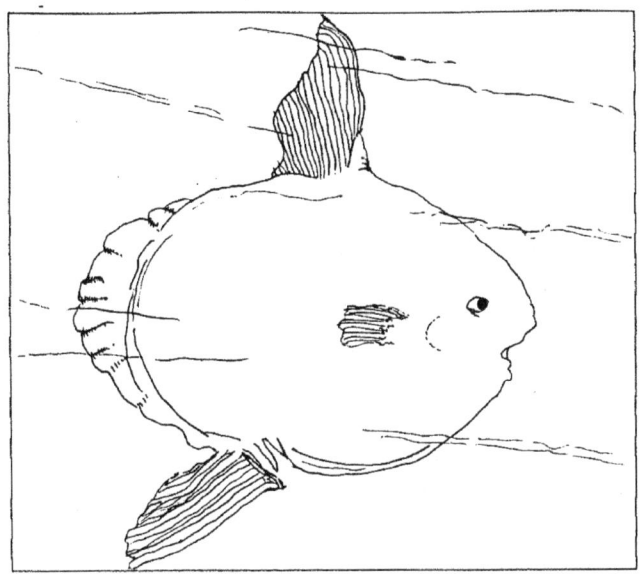

Sunfish *(Mola mola)*

The final report that I have uncovered is not really about a fish at all. It concerns a whale. The discovery of the whale in question was reported in the same edition of the Fleetwood Chronicle that covered the landing of the aforementioned Sunfish. The partially decomposing whale was about twenty feet (6m) in length and the weight was estimated at between two and three tons (2032-3048 kgs.) The carcase had been washed up at Rossall point and after failing to move it, the men from the local Highways department decided to bury it where it lay. You may be forgiven for thinking that the discovery of a beached whale has nothing to do with cryptozoology. Up to a point you would be correct in thinking so, but I think that it demonstrates that the most startling discoveries are to be found under our noses. For instance, I know the area of beach around Rossall point very well, as it is only a couple of miles from my home, but I had no idea that beneath my feet while walking on the sands lay the skeleton of an immense sea creature.

REFERENCES

Fleetwood Express 17th July 1897 p.21.

MAITLAND, P.S & CAMPBELL, R.N., *'Freshwater Fishes'* (Harper Collins pub, 1st ed. 1992, pp 333)

Fleetwood Express, 6th May 1905, p.8.

Fleetwood Express, 3rd June 1905, p.8.

Fleetwood Chronicle, 19th September 1913, p.8.

Fleetwood Chronicle, 18th May 1961, p.1.

BRIGHT, M. *'There are Giants in the Sea'*. (Guild Publishing, 1st Ed., 1989, pp 185-6).

Fleetwood Chronicle, 2nd September 1965, p.17.

EDITORS NOTE.

The descriptions of the sharks and the drawings were adapted from *'The Jaws of Death'* by Xavier Maninguet. (Harper Collins 1992).

* * * * *

The Beasts of Kingsteignton.
An investigation by the Centre for Fortean Zoology.

The CFZ research-team on this project consisted of John Jacques, Jonathan Downes, Graham Inglis and Alison Downes.

Introduction.

Andrew G is a man of late middle age living in Kingsteignton, a large village which forms a suburb of Newton Abbot in mid-Devon. He had been in contact with Mrs Jan Williams during the period when she was running S.C.A.N. When Mrs Williams joined the C.F.Z she suggested that as the C.F.Z was based in Devon, some researchers from the C.F.Z should approach Mr G for some further investigations.

This we did in the early spring of 1995.

MAPS OF NEWTON ABBOT AND KINGSTEIGNTON.

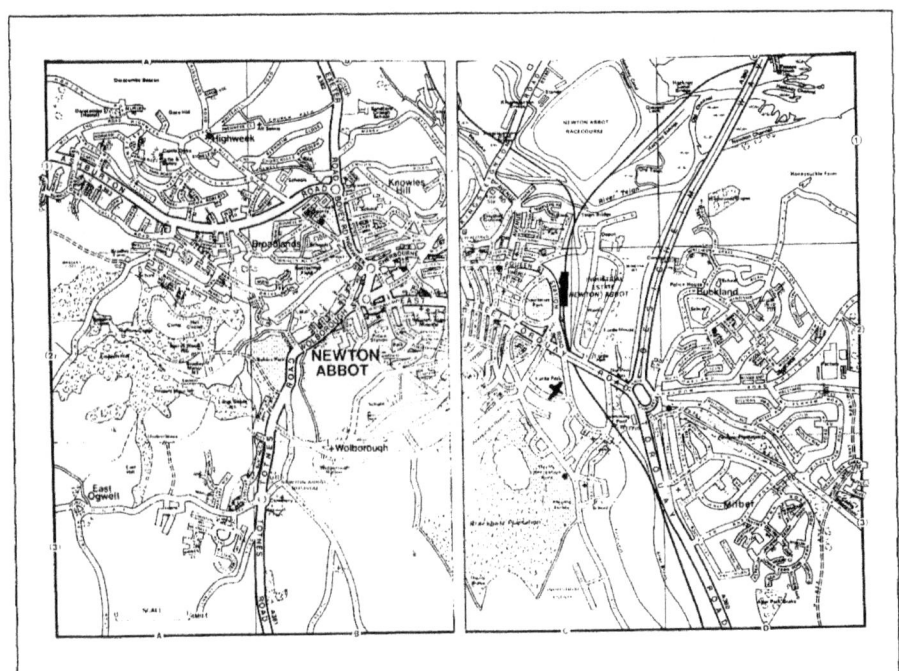

NEWTON ABBOT

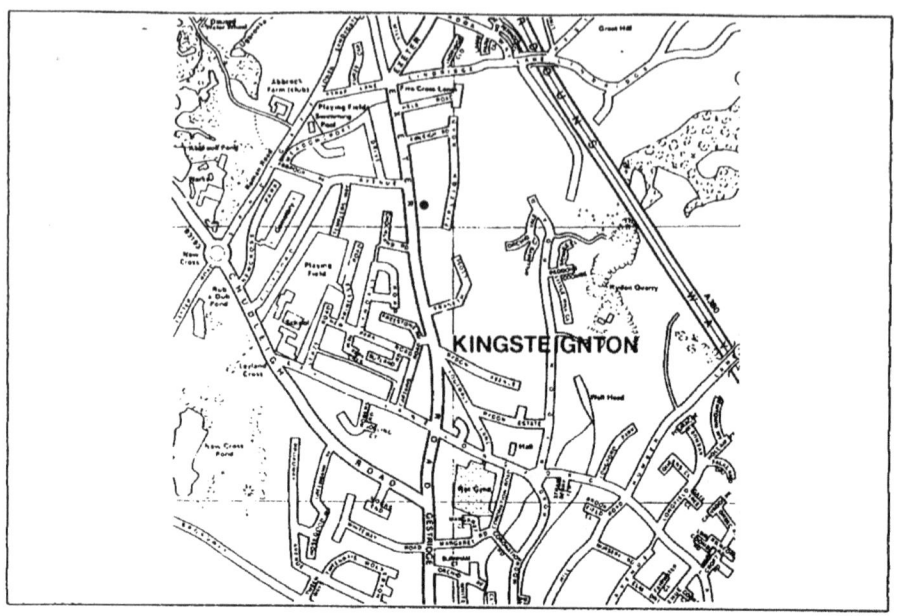

KINGSTEIGNTON

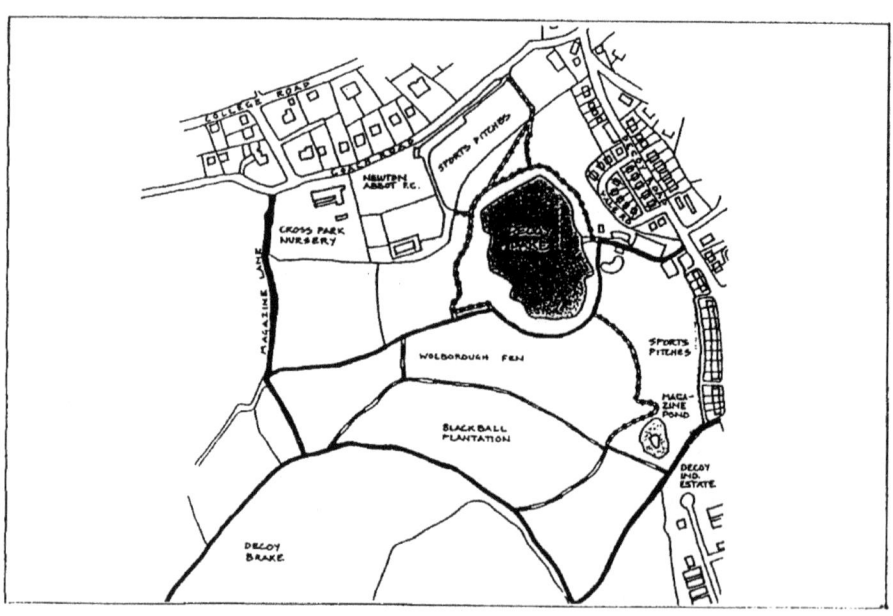

DECOY COUNTRY PARK

At the time this investigation took place, Decoy Park, the site for many of these sightings was an unexciting looking council run leisure ground surrounding a lake which had, apparently been created from an old gravel pit. The woods on the far side of the lake included an area of restricted access that functioned as a nature reserve, and it is well known that badgers were breeding in the area.

TEXT OF THE INTERVIEW CARRIED OUT BY THE CFZ RESEARCH TEAM:

A.G: Andrew G. (Interviewee).
J.J: John Jacques. (Counsellor).
J.D: Jonathan Downes. (Observer).
G.I: Graham Inglis (Cameraman).
A.D: Alison Downes. (Observer).

(NOTE: The transcript of the video is presented almost exactly as it was recorded. Ums, ahs, polite small talk about cake, and bits where several people were talking about nothing in particular at the same time have been edited out in the interests of readability. The purpose of this research paper is, however to present all of the available facts rather than to provide a concise overview and our original video testimony is available for scrutiny).

G.I: If you could speak up for the microphone and ignore the camera...

J.J: Yeah. *(To Andrew G)*, so would you like to tell us in your own words what you have seen?

A.G: Well the first cat I saw was standing on my garden wall about eight feet from the front door at midnight. The dog started barking at the door. I opened the front door, and there on my garden wall was a large pair of back legs and a tail. Andy coloured. I couldn't see the front. It must have been standing on the bath looking over the other side of the wall. The bath was three foot six from the top of the wall. Estimated size of cat five foot six. That was the first one...and that was in '85? yeah '85.

The next one was a black one in Penn Inn park. That's where they've now built a supermarket...

J.J: Penn Inn Park. Was this like a piece of rough ground?

A.G: No it was a proper p a r k . Swings, paddling pool, rabbits. That is why we used to go down there, because the dog loved chasing the rabbits. One night I just followed this large, five foot six, black cat into the park. Then it disappeared and I didn't see it again until the following year....this was on the other side of the railway.

This is the secret really. The railway. When it came into Penn Inn park it came off the railway. Then I discovered they were coming through my garden. This is how I saw the first one.

J.J: So it gets to the railway line beyond your garden?

A.G: Penn Inn all used to be a park. Its now a supermarket, but the cat came down through the gardens, across the road and into the park. There is a tunnel where the road goes under the railway. The second time I saw it, it jumped across the road and into another garden. Which ties in nicely with my first sighting on my front lawn. They were using my front garden to get from the lake to the park. When they turned the park into a supermarket I had to find another park for the dog. So did the cats.

THE BLACK CAT DRAWN BY ANDREW G.

J.J: They'd been disturbed?

A.G: They'd been disturbed. So now they had to go round the back way to the lake.

J.J: They are actually living on the outskirts of town?

A.G: Yeah. They actually go through the suburbs at night. They go right across Forde Park to get there because they won't go across open ground.

I'd seen this puma previously across the park from my front room window.

There used to be a fox which used to worm on the grass and so I would go out with my binoculars and watch this fox. If I wasn't careful it would see me and it would be gone. This puma was hunting in the park. The next time I saw it was Christmas 1989/90. Three sightings. One before Christmas, the second one after Christmas. It walked across the road real sneaky. You know how they do these leopard brooches. It was slinking.

J.J: Are you fairly sure that it was the same one?

A.G: I don't doubt that it was the same one. I reckon that it stayed in the area for about a month, two weeks before, two weeks after christmas...something like that. The third time I saw it, I just happened to walk past this garden. I looked up this path and there it was walking towards me down the path. I stepped back and it kept on walking towards me. It totally ignored me. It came within 25 feet and turned left into the garden next to it. Some time later someone in Teignmouth spotted a big cat and I reckon it was the same one.

J.J: By the sound of it, this particular one wasn't too worried about human beings was it?

A.G: I reckon it tried to have a go at me prior to that, but being out walking every night it got used to me. I saw it four...no three times, but it probably saw me a dozen times..

J.J: So it was used to you.

A.G: That's right. I worked out then that this is how they were getting from the corner of the railway right across to the lake. That took me ten, no six years!

The black and white one. I saw it at the lake first. This was the ONLY one that I saw in daylight. It was 07.15, a nice, bright, sunny morning. Just over here is a little sailing club. I was standing by the sailing club when I saw this white spot dancing in the trees. And I was watching it, and then it appeared on the other side of the lake, and it was obviously a big cat. It walked down the edge of the lake, in hunting mode, down low, and it came to a bank, like a cliff, taller than me at any rate, and it just got up on its hind legs and looked over it. I reckon it was seven foot. I estimated it at first at six foot, but it must have been taller than that. It tried to take a peek over the bank, but couldn't see because the bank is round, so it just shot its head up another foot over the top and looked over.

J.J: You're saying that its white?

A.G: Its all white except for the.... at that time it had a black band around its neck, a huge patch on the back like a blanket and a white and yellow banded tail. The whole cat was fluffy it had huge paws....

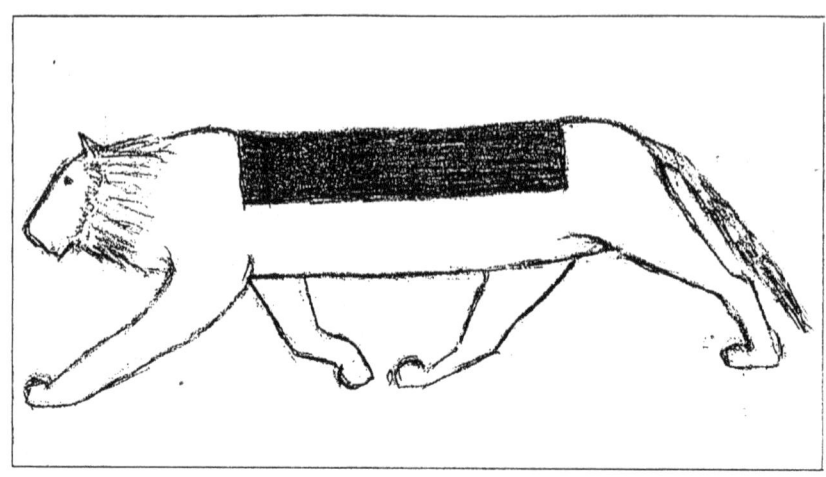

THE BLACK AND WHITE CAT (First Sighting)
by Andrew G.

J.J: The patch on the back was this square shape?

A.G: If you think of tigers. Tigers have stripes don't they? Its all straight lines isn't it? Thats what this patch is ... straight lines. When it stood up on its hind legs I could see that the sun was shining through the fur and I could see the skull. There was two inch fur on the head.

J.J: So you've seen this white animal again?

A.G: I saw it again the following year but this was at night. Crossing the road at Forde Park. Out there across the field. Three quarters of a mile it came walking out of the woods. You think, 'is it a dog?', but it's too big for a dog, and I couldn't think of what animal it could be so I thought it was a person. All dressed in white wearing a white hat. I could see the leg movements and they were just right for a person. I thought that it was going to be behind the hedge on the near side soon if I didn't get a look at it, so I got the binoculars, and then it turned sideways and I recognised it instantly and there was no mistaking it.

J.J: So you were actually in this room and you saw it over there through binoculars.

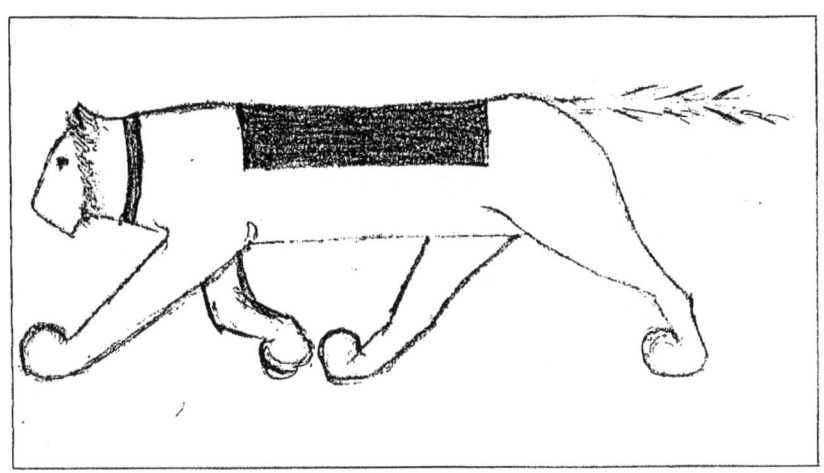

THE BLACK AND WHITE CAT (Later Sighting)
by Andrew G.

A.G: That's right.

J.J: That was when? Day time? Night Time?

A.G: Three o'clock in the afternoon. Nice sunny day. It shone. With the sun on it it looked beautiful. It stood out like a sore thumb. Obviously I'd never seen anything at that distance before so I didn't know the scale. It was too big for anything else so I thought it was a person. A short person, maybe five feet tall, wearing a hat. In February *(laughs)*

J.D: Have you seen any footprints?

A.G: The biggest paw-print I ever measured was five and a half inches. That was at Decoy lake. There was white fur there on a barbed wire fence. I saw some sizeable footprints. Some left by the black one-I reckon that's the 'Beast of Exmoor'. The 'Beast of Exmoor' is supposed to have three toed paw and I've measured a three toed paw at four and three quarter inches. That makes it a fair old sized cat.

J.J: These two drawings here appear to be of the same animal. Except for that (points to band round neck)...

A.G: It must have had a collar on and now it's fallen off. I saw..I thought I saw, a line on the neck that was the colour of flowerpots. The er.. shiny light brown. When I first drew this I got the head wrong. Jan Williams questioned me on this, so I had to accelerate the binoculars so that I could see the head. It was lion shaped.

If I had only seen the head I would have said '*that cat-it was a lion*', but the rest of it...'its a tiger'.

J.J: If these sightings are years apart, then the creatures could grow in size, grow manes, whatever..

A.G: Exactly!

...4th May 1991. The big black cat. It was scent marking the tree. I must have been about eighty yards from it. We spotted each other at about the same moment. It was dark, but the cat stood out because the street lights were behind it. It was in silhouette and it was big. It saw me, and I saw it, and it jumped straight up to the top of the park in about four jumps.

The park is on a slope. I decided to walk up to the top of the park and see if I could see it again.

J.J: Were you sure it was black if you only saw it in silhouette?

A.G: Oh yes. The park, is, I suppose about two hundred yards across at the top. Up the slope, up in front of me, lying on the grass I could see a coconut.. It looked like a coconut with the pointed end upwards. I thought 'this is strange', because visiting the park every night you get used to what is in it. I looked around for the dog but he was miles behind, and I took three steps to the right and this 'coconut' got off the ground and did 35 mph down the road and disappeared into a garden gateway. It was so fast you just get an impression of size and speed. As it turned the corner into the gateway the back end bounced as if it were braking! (laughs)

J.J: Has your dog ever reacted to these animals?

A.G: Before I ever saw one. This was in '84 when I first became aware. I was doggy walking and there was a report of a puma seen at Ogwell, and doggy walking around the lake I thought I'd keep my eyes open. The first morning there was a paw print. No claws. I was interested. A bit further on into the wood and there is a tremendous hum of flies like an aeroplane engine. The wood is on a hill with all ferns six foot tall. I noted that. I wasn't sure what we were up against and you don't go diving in. That was one day.

Then I noticed the scent marking. If you ever went into the lion house at the old London Zoo then you'll know what it was like. I used to go there all the time as a child. It was a shilling a day. I'd sent my wife down to Torquay for Di Francis' book 'Cat Country', at the time. I think Di Francis had heard about the sightings of a big cat in Decoy Woods. She was in there one day, standing beside a tree that absolutely ponged. I didn't know it was Di Francis at the time. I only recognised her from photos later. I could have shown her a pawprint.

Later on Toby, suddenly shot into the ferns and went rushing up the hill. I stood and I shouted. Then he came tearing down past me. He didn't stop. I don't know what he discovered but it put the fear of God into him. He never left my side again.

JJ: Can I ask you something? When you've been telling me about this, you've made a lot of estimates about distance. Have you had experience in this kind of thing?

AG: I was a joiner. I'm retired sick now, unfortunately, with emphysema ..... I spent a life time measuring with a tape measure in my hand. I'm reasonably sure of dimensions.

Scale is the problem. I understand scale. The little puma crossing the road. How was I sure of its size? It was five foot - the width of the pavement plus a foot. It didn't jump the gate so the gate must have been open.

The white animal)...when I told you where it stood it filled the slope. When you saw it from a distance it seemed smaller.....and then there's one that is similar to a puma but with no markings. That's the sandy one - my original cat. I saw it jump the road in February '91. It completely cleared the road at Forde Park. There have been four sightings of the big cats at Forde Park.

J.J: You've seen three animals then. The sandy coloured one, the big black one and the black and white one.

A.G. Yes, and a standard puma. That was sandy coloured again. Obviously puma because of the markings. The other one looked like a puma but had no markings. All pumas have a black tipped tail.

J.J: I'm sorry that I'm asking you what may seem like obvious questions, but I want to hear the answers in your own words...

A.G: Just so. Quite so. The one that looks reasonably like a puma but isn't sort of thing. It's a hybrid puma, let's put it like that. It jumped twenty foot across the road. Then it did another jump of twenty foot, and then a ten foot jump and then down to a walk - our sort of speed - in one pace. I walked away from that one. I could have seen more, but I didn't realise that they were crossing the park until later in the year.

J.J: This lake. Are there parts of it that are overgrown?

A.G: Oh yeah, except now they've got wardens and they are civilising it. It is still very wildernessy. I've seen a cat that's about fourteen inches long cross our path one day. Then there was a grey one, about twice the size of a domestic cat...they've had all the rabbits in the area and nearly all the squirrels, and the council has destroyed what's left of the area.

* * * * *

ANDREW G'S WRITTEN STATEMENTS.

(These are taken verbatim from two pages of handwritten notes given to the CFZ research team by Andrew G on the day we first interviewed him).

July 1984. Time 2.30 AM.

The roar of the big cat stopped me dead in my tracks at my own front gate. I turned to face the direction of the sound. This was a full round, lion-like roar the awesome power of which forced me back two steps, and I half turned ready to run. The roar lasted for about 8 seconds dying away at a higher pitch than it started.

Then came the deep hiss lasting the same amount of time as the roar, coming down on top of me and totally surrounding me.. At this point my brain stopped working. This is England, it just doesn't happen I thought.

From that night onwards I've known that out there in the countryside are lions or tigers or both.

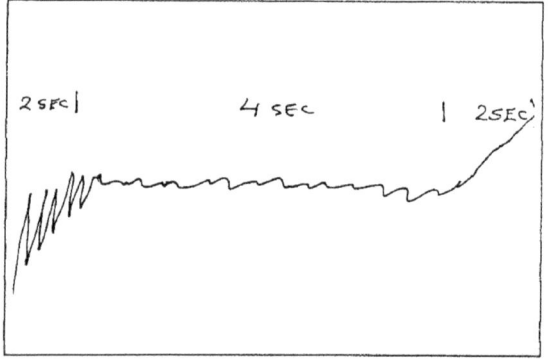

Schematic depicting 'tone' of the roar heard by Mr G.

Puma seen, Ogwell 1984.

DECOY August 1984.

South end lake. Paw print. Claws retracted. Large, approx 3.5 to 4 inches.
Scent marking.
Hum of flies in woods.
Di Francis in wood?
Roar of lion finishing with hiss 2-2.30 AM.

Toby dog followed scent.

November 1984.

Found bones where flies were buzzing.

August 1985.

North end lake. Paw print. Claws out 4 inches. Midnight back legs and tail of cat on garden wall. Rest of cat hanging over wall on edge of bath. Bath to top of wall 42 inches. Cat light sandy colour. Short hair. Long thin tail. No markings. Rough estimate - length of cat five feet to five and a half feet overall.

October 1986.

Black cat crossing road at Penn Inn Park. Length five feet to five and a half feet overall. Followed it into park but didn't see it.

March 1987.

Black cat again crossed road very fast. Keyberry Park. Dogs barking.

March 27th 1988.

7.15 am. Sunny morning. Black and white cat, other side of lake. Length 6 feet to 7 feet overall. Like giant persian.

1989/90 Christmas.

3 sightings puma. Length five feet.

1991 January.

Growls from garden.

1991 February.

Forde Park. Lion jumped road.

1991 May 4th

Black cat scent marking tree. Forde Park, could be seven ft. Head and body only, then lying on grass watching me. It did 90 yards in 5 secs. at a fast walk. Kingsteignton, higher Sandygate.

Feb 3rd.

Black and white cat again. Now 8-9 feet head and body. Looked like type of tiger.

Feb 17th.

Black and white cat returns.

> (NOTES: It has to be noted that the events recounted by Andrew G in our interview and the events written down in his handwritten notes do not tally exactly. Andrew G told us off camera that these events had come at a stressful time in his life, and it seems likely that his memory of events has become a little distorted. In the opinion of the researchers, a certain degree of leeway is a favourable piece of evidence as it tends to indicate that the interviewee's account is not a well learned 'party piece' recited 'parrot fashion' to interested parties).

We returned to see Andrew G again in May 1995. This time we were accompanied by a TV crew, and interviewer Ruth Langsford. A six minute item was eventually included in the CST Productions series 'Mysterious West' in October 1995.

TEXT OF INTERVIEW WITH ANDREW G ON 'MYSTERIOUS WEST' TV SHOW:

> (The interviews with both Andrew G and Jonathan Downes were filmed by CST on the shores of Decoy lake. The Andrew G interview only is printed below):

A.G: February 1988 a white patch appeared in the woods over there coming towards me. I watched this white patch and it appeared as a big cat, on the edge of a cliff over the other side there. It turned right and went along the top edge of the cliff and along the path over there, and there's like a sort of small cliff over there. The cat just stood on it's hind legs and looked over the top of the cliff. Its about six foot high! It just took a peek at first and then put its head up another foot and then looked over the top. Then it climbed on top and surveyed the whole area. It looked at me, and I was looking at it through binoculars, and a shiver went through me, and I left it standing there and I carried on my walking the dog around the lake.

R.L: When you first saw it, what did you think when you first when you first saw something white. What did you think then?

A.G: Well, it looked so peculiar. Like a ghost dancing in the trees. Something white, flickering. Then when it came into the open it was a cat.

R.L: When you say a cat..obviously not a domestic cat.

A.G: No no no no no. I estimated going from what it was looking over down there it was seven feet long. Black and white with a white and yellow banded tail and a black band around the neck. Very fluffy.

R.L: Did you recognise it as anything. A puma or something?

A.G: Nothing at all. I couldn't put anything to it...

R.L: What do you think it could be. If it's like nothing you've ever seen before.

A.G: Its like a Siberian Tiger but the colours are all wrong. You can have a black and white tiger I know that, you can have a white tiger, but this one is so different.

R.L: Why do you think they are here?

A.G: Well, they like to be near water. This is their route round the town.

R.L: You think they go wandering in Kingsteignton and Newton Abbot?

A.G: They don't go wandering around Newton Abbot itself, but they go around town on the railway.

R.L: *(Obviously misunderstanding and having visions of these huge creatures politely queuing up for platform tickets)*, Why do you think that?

A.G: I've seen them near the railway too often.

R.L: Do they come through the park?

A.G: They come through the park, up through the gardens, around the station and away into the countryside on the other side of town.

R.L: What sort of reaction have you had from people when you've told them about this?

A.G: The wife doesn't believe me of course, but then I keep on. You've got to believe what you see.

R.L: Does that worry you that people will think that you've gone a bit crazy?

A.G: You get used to it. People either avoid you or don't talk to you. They think you're mad as you say....

(NOTES: Andrew G's second account was similar to the first and the minor discrepancies can, I think, be put down to nervousness at being confronted with a glamourous TV presenter and a TV crew).

CONCLUSIONS.

This is a complicated case. It is too early to reach any firm results but some interim conclusions can be extrapolated.

Firstly, it is the opinion of the CFZ investigating team that Andrew G is sincere in that he believes that what he says is true. He believes that he did see the animals he describes at the times and places he lists. He was, however, at the time under a great deal of stress in his personal life, and has more recently ceased working because of illness.

Andrew G admits that he has been a cat afficianado for many years. He talked in the interview about spending all day in the London Zoo Lion House, as a child. His house has many beautifully illustrated books on cats of all sizes and shapes. He is unarguably a fan.

Unfortunately, checks around Newton Abbot/Kingsteignton have uncovered no corroborative witnesses to the giant black and white creatures or to the 'sandy' animal. There have been puma sightings and sightings of the black creature both before and after the dates mentioned by Andrew G.

The grey cat 'about twice the size of a domestic moggy' has been seen in several locations in Newton Abbot, Kingsteignton, Ipplepen and the surrounding areas. Indeed there have been so many sightings of this, or similar creatures, that one would suggest that a sizeable population may exist.

The giant black and white creatures, which Andrew G alone has seen do, however pose a far greater enigma. The hair samples he provided were analysed by Dr Kitchener of the Royal Museums of Scotland, but the results were disappointing. (For full details of the analysis see A&M3).

It would be tempting to dismiss these sightings as delusions or hallucinations brought on by stress or even wishful thinking rather than as true subjects for investigation. The recent sightings of similar creatures in Devonport and Saltash, however must ensure that, for the time being at least this case remains open.

In the TV show Jonathan Downes discussed the possibility that these animals, which have certain characteristics reminiscent of mustelids rather than cats were actually an evolved descendant of the cave Wolverine, but even then we admitted that this hypothesis was highly unlikely. It seems likely that when Andrew G describes a 'white ghost dancing in the trees', he may not be so far from the truth, and that whereas the sandy, black and grey cats he has seen have a corporeal reality, the awesome, black and white beasts of Kingsteignton have a far less tangible nature.

* * * * * * *

About the survival of relict hominoids from the point of view of a Zoologist.

by Francois de Sarre

When considering the possible survival, to the present day, of wild, hairy men in various locations on the planet, the evolutive model that is currently proposed, seems not to be providing wholly reliable answers.

It would seem to be quite the opposite. If we are to accept the conventional belief that man descended from the apes, a certain evolutionary logic would imply that the man of 'modern' type, and that the 'primitive forms, such as the hairy wild hominians would have been completely supplanted.

The consequence of this way of perceiving the facts, which has prevailed in science until now, was to discredit 'Cryptoanthropology', or the science of hidden men. A particularly revealing example was the reaction of the anthropological establishment during the affair of the Minnesota Ice Man (HEUVELMANS 1974).

Nowadays such French palaentologists as Yves COPPENS or Jean CHALINE (see in: GRISON 1990), are more conciliative and are prepared to consider the possibility of the survival of Gigantopithecus in the well known form of the yeti.

This however, is not directly relevant as it is commonly accepted that Gigantopithecus is descended from some Miocene dryopithecines (COPPENS 1983), and is a dead end, out of the direct line of human descent.

Another problematical concept for the palaeontologist is the possible survival of neanderthals like the Caucasian 'Almasty' and the 'Bar-manu' of North-Pakistan. (KOFFMANN 1991, MAGRANER 1992). The scientists I cited earlier believe that palaeontologically Homo sapiens neanderthalis is very close to modern man.

There is in fact no scientific doubt that men of the 'modern' type and Neanderthal-men lived together for at least 100, 000 years during both warm and cold periods. (VANDERMEERSCH 1988, BAHAIN et al. 1993). We only ask why this state of affairs could not have continued to the present day? Picking up where he left off in his previous paper on this subject, (de SARRE 1991), the author of this present paper reaffirms his personal view that Hominoids of diverse types, have co-existed with man for many million years. Their co-existence in contemporary time would therefore be a logical continuation of an ancient biological state.

The Simian Anthropomorphs

*All the way through this article we refer to the checklist
of the world's hidden animals compiled by Bernard HEUVELMANS
in 1986.*

The Primates are generally divided into three sub-orders; the Prosimia, the Tarsia and the Anthropoidea. The last group contains the New World Platyrhinea and the Old World Cararhinia, which themselves subdivide into the Cynomorphia (macaques, baboons etc) and the Anthropomorpha (people, australopithecines, chimpanzees etc).

The Simian Anthropomorphs are the tail-less apes, both recent and fossil. They are characterised by their aptitude for quadrupedal locomotion. Within this group we find the most representative of all Cryptozoological 'monsters', the 'yeti' of Nepal.

Fig.1 Classical Yeti (Dinanthropoides nivalis) after Heuvelmans (1993), V.S.D/Nature 5;69.
Picture Courtesy Francois de Sarre

Scientifically described by Bernard HEUVELMANS in 1958, under the name *Dinanthropoides nivalis*, the so called 'abominable snowman', is in reality a big ape, very similar to the African gorilla which, itself, was a semi mythical 'monster' until the year 1847 when it finally became officially recognised by science.

The yeti (who we should really call the mi-teh), like the gorilla in the last century, is often exaggerated in size with the aim of striking the imagination. Only its reputation is 'abominable'. As Bernard HEUVELMANS (1993) noticed many years ago, the figure of the yeti is used to frighten sherpa children when they don't eat their dinner as a Nepalese equivalent of our bogy-man.

As a zoological creature this big ape leaves its curious four toed tracks in the snow as he walks, occasionally bipedally, through the Himalayan passes. It usually dwells amongst the rhododendrons, the bamboo and the birch trees, in the forested valleys of Nepal at an altitude of about 4,000 metres. (HUTCHINSON 1991). The form of its skull, which resembles a 'sugar loaf' reveals the presence of an osseus sagittal crest which is covered with scrubby hairs. A similar formation is found in the male gorilla, and the robust australopithecus which suggests that the yeti may have developed heavy 'grinding' jaws to deal with its tough vegetarian diet.

One characteristic of the yeti which is also a parallel with the gorilla are the formidable fangs that it shows to frighten people. Contrary to what many people believe, these huge simian teeth have nothing to do with carnivorous habits. They instead result from a forced adjustment of the jaws as the permanent quadrupedal gait became effective. The development of teeth in the primates works with the lengthening of the jaws. (FRECHKOP 1940). Imagine a human being on all fours. he has to lift his head up in an exagerrated manner in order to see in front of him. In a quadrupedal animal, the spinal chord has to be inserted higher in the back of the skull whose structure then becomes deeply modified. In compensation, the snout grows well. (WESTENHOFER 1935, HEUVELMANS 1954). Further developments lead to primates with very long muzzles like baboons or lemurs.

In the hypothesis which suggests that an initial bipedalism was the primordial gait amongst all primates, the original form of the foot was plantigrade, like in humans. In apes, however the passage from a terrestial life to an arboreal one, from a bipedal locomotion to a quadrupedal locomotion converts the foot into a posterior hand with an opposible big toe. The footprints of the yeti are characteristic of an ape who kept this big toe, as opposed to the footprints of the sasquatch, for example, which are of a human type.

We can see the yeti as an animal, which passably resembles the mountain gorilla and which also lives in high altitude forests. Like the gorilla it is faced with extinction and it would be a real pity if it vanished from the Himalayas before the scientific world got around to admitting that it had ever existed.

The Australopithecines.

In Africa there are still persistent rumours of hairy bipeds which appear to be surviving

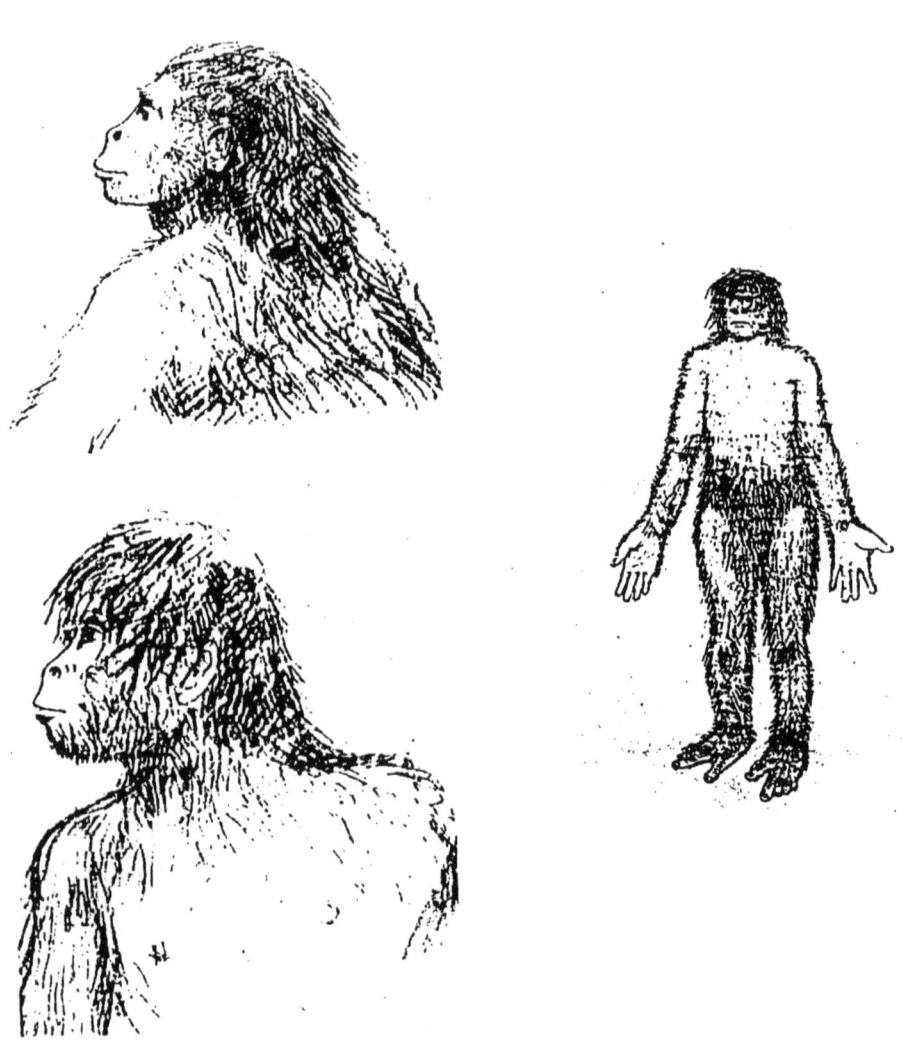

Fig.2. Representations of the 'Kakundakari' (left) and of the 'Kikomba' (right). After Heuvelmans (1980 46-7)
Picture courtesy Francois de Sarre

australopithecines. This would not be that extraordinary, at least from a zoological point of view.

We know the story of the palaeontologist John T. ROBINSON who acknowleged that he had set traps in the secret hope of catching a living australopithecine.

Known by various names, (Kakundakari, Kikomba, Agogwe...), these creatures are described as quite bipedal. This agrees with what we know of the fossil australopithecines, even if they had developed a real tendency to climbing up trees. Witnesses also speak of 'long head hair', and say that the canines 'do not over-range the other teeth'.

There is little doubt that some of these wood and savannah creatures are australopithecines of the gracile type. (HEUVELMANS 1980). Their survival to the present day is as simple to explain as that of the african apes, with which they do not compete. The biggest threat to them is to be found in the destruction of their environment and in the present proliferation of another kind of primate-Homo sapiens!

Let us here, also mention the work of French Ethnologist Jacqueline ROUMEGUERE-EBERHARDT (1990) who studied the unidentified hominoids of Kenya. Some of the observations be could be of australopithecines or pithecanthropes. It is quite difficult to make the distinction and we do not really know what these creatures, that the palaeontologists exhume, actually looked like. The pithecanthropes had larger bones and skeleton, and their superciliary arches were more salient than those of the australopithecines.

The author believes that these diverse lineages were the result of successive evolutive waves branching out from from the ancestral trunk of human ascent. (See Figure 4).

There has been a lot of media coverage of 'Lucy' *(Australopithecus afarensis)* and her 'son', and one should remember that on the same location in Hadar, Homo type fossils were also found. At 3.5 million years old these are a lot older than both 'Lucy', (who has often been referred to as 'the mother of humanity'), and the new skull discovered by William Kimbel and Donald Johanson.

The article in Nature 368, from the 31st March 1994 makes no mention of the human fossils in the stratigraphic scales of Hadar. Is there a real campaign of disinformation around the world, by people who keep insisting that australopithecines pre-dated man?

The Giant Hominids.

Under this name we shall include those giant (over three metres tall), hairy bipeds whose footprints resemble our own, with the only difference being that they are excessively large. Bigfoot! They are well known through the plaster casts of the prints of their bare feet. They are also notorious for numerous hoaxes and mystifications like the Patterson film. The North American Sasquatch was scientifically described by Grover S.KRANTZ (1986), of Washington State University, as Gigantanthropus canadensis, after examination of footprint casts which showed dermatoglyphs.

The German palaeontologist Franz WEIDENREICH (1945) already suggested this generic name as a replacement for Gigantopithecus, which was discovered in 1935 by the Dutch naturalist G.H.R. von Koenigswald, who discovered an enormous tooth for sale in the shop of a chinese apocothery.

Currently, a few mandibles and hundreds of teeth of Gigantopithecus have been discovered. In hand-books it is generally represented in a quadrupedal position, like a sort of huge gorilla (COPPENS 1984), usually leaning on the back edge of his bended finger phalanges. (Knuckle walking).

In 1952 Bernard HEUVELMANS suggested the possibility that Gigantopithecus could be the identity of the largest yeti. Both HEUVELMANS and Grover S.KRANTZ believe that there is a close relationship with the Sasquatch, because the canines of the fossil Gigantopithecus are known to be very reduced, and the whole dentition seems to be rather more human than ape-like.

Furthermore, if subsequent discoveries confirm this viewpoint, and consign both the fossil primate and the giant wild man to the same genus, it will be appropriate to use the designation Gigantanthropus for both, (the zoological name taking precedence over the palaeontological name), finally granting Franz Weidenreich's wishes.

The Asiatic 'Nyalmo' (also called 'large yeti' by westerners), is present in the south of Tibet, in Sikkim, Birman, China and in peninsula Malaysia where it is known by the local name of 'Jarang gigi' (teeth set at intervals').

As he returned from a period of study and travel in 1993, Dr Heuvelmans told me the enjoyable story of Captain Mokhtar Mohamad. In the Malaysian forest, he suddenly felt the touch of an enormous, hairy hand upon his shoulder, and looking back, he saw with great fear a three metre tall 'Jarang gigi' standing behind him.

These giant hairy men are quite peaceful and have the best intentions towards us. Perhaps they even try to come into contact with people. We must hope that this friendly behaviour does not provide the occasion for carnage, under cover of science, as they are trailed by a pack of head hunting killers. Fortunately for the Sasquatch, in the USA such hunts have been fruitless.

According to witnesses, the giant, hairy men of Malaysia have a long fair mane of hair which falls upon their shoulders. At the bottom of the back they have a sort of little tail that gives the impression of being rather 'simian' or 'beastly', but this author believes that it is in fact a mark of the archaic human lineage.

My reconstructions of the appearance of the bipedal homonculus who was at the origin of the mammals endow our remote ancestor with such a caudal appendix, probably the inheritance from a former aquatic stage. (de SARRE 1989, de SARRE 1992). Among contemporary man this atavistic feature sometimes reappears. (SIEVEKING 1989).

Fig. 3. Representations of the 'Bar-Manu' (left) and of the specimen of Homo pongoides (right). After Magrainer (1991, p.13) and Heuvelmans (1993 3º Millenaire, 28;55)

The relict Neanderthalians.

Asia contains another extraordinary being; one so close to us through its general anatomy that the majority of scientists still class us as the same species. *Homo sapiens (subspecies neanderthalis)*. The man of neanderthal.

Everyone knows the prehistoric figure that science has always presented to us as our 'ancestor'. Now, however, we know that men of the 'modern' type preceeded the neanderthals in the near east. It may even be, that in Europe, during the last climatic pertubations, the neanderthal subspecies who were better adapted to extreme cold unequivocally supplanted the local H.s.sapiens.

Purely on the basis of the lack of officially recognised remains, more recent than the thirty thousand year old skull of Saint-Cezaire the palaeontologists have sought to bury the man of neanderthal 'with the honour due to him'. Michel RAYNAL (1994) noticed that behind the dogma of the extinction of the neanderthal men, is hidden the idea that they have been 'metamorphasised' into modern men.

Logically, as the climate became milder in Europe, the neanderthals had plenty of time to take refuge in the large mountain massifs such as the Pyrenees (RAYNAL 1989, RAYNAL 1993), in Caucasia (KOFFMAN 1991), in Pakistan (MAGRANER 1991), and elsewhere in Eurasia (HEUVELMANS 1993).

The deep frozen specimen that was examined by Bernard Heuvelmans, and has become known as 'The Minnesota Iceman' probably originated in Viet-nam. (HEUVELMANS 1974). Perhaps it will soon be possible to observe a living, flesh and blood neanderthalian, thanks to the efforts of a naturalist called Jordi Magraner who is studying the 'Bar-manu' in the mountains of Pakistan.

The original features of the neanderthalians, fossil, or contemporary should be considered as adaptations to a cold, mountainous environment. The fore-arm and legs are relatively short and the facial sinuses are considerably developed.

As the American researchers E.Trinkaus and F.Smith have suggested, most of the cranial and mandibular characteristics of the Man of Neanderthal are related to an intensive utilisation of the anterior teeth for non-masticatory employment. The mouth became a sort of third hand with significant consequences for the morphology of the face (oncognathy) and for the reshaping of the hind skull.

We can see a weak opposibility of the thumb, with consequently less aptitude for manual ability. This can be explained by the utilisation of the mouth for seizing. The thumb, composed of two phalanges of equal length, could, in compensation ensure an extremely strong grasp. It is said that the colossal Neanderthalian hand was ten to twelve times more powerful than our own.

The feet of the Neanderthalians are very large, with curved, movable toes arranged in a fan shape in order to facilitate grip on rocks whilst climbing. In comparison with *Homo sapiens*, who is diurnal, gregarious and mostly a plains dweller, *Homo Pongoides* [*] is a being who lives alone or in pairs in high mountains and remote areas with essentially nocturnal habits.

[*] We should really call the Man of Neanderthal by his Zoological name, and no longer with his palaeontological one. The subspecies neanderthalis now falls into caducity.

Discussion.

The objective study of the palaeontological facts, and the results of the Cryptoanthropological research, produces evidence which suggests the following:

1. That the survival of cryptic wild and hairy hominoids to the present day is not in itself any more extraordinary than the survival of the diurnal great apes in the relict forests.

2. That a great deal of Cryptoanthropological field research has been obscured by tenacious evolutionistic prejudices in favour of the recent emergence of 'modern man' from ape stock.

3. The fossils collected indicate that there were periods during which the non sapiens hominoids were significantly widespread during a period of regression for the human species. One example being during the last ice-ages.

4. That since remote geological times there has been a co-existence between man of 'modern' type and those wild, hairy hominians, who are dehumanized but also descended from common ancestors.

Figure four (see page 107) shows a bushy evolutive family-tree of the human species, including post-human forms (that I call hyperanthropoids), known as fossil hominians or as 'relict' hominoids that have survived into our present time just as
man did.

On the left side is the representation of the traditional conception of the palaeontologists, in a frame that leads a supposed evolution from a pre-australopithecine to modern man. The arrows allow us to pass from one side of the table to the other, and to compare both conceptions.

The phenomenon of dehumanization, as described by HEUVELMANS (1974), in the midst of all the hominoid lineages associates a cultural decline with anatomical changes:

"The forehead recedes, the jaws are developing. the masticatory apparatus becomes mightier which induces an enlarging of the osseous crests of the skull to which the concerned muscles are linked. The whole silhouette can modify itself, the head sinks into the shoulders, the body is more and more inclined forward, tending to horizontality and a quadrupedal gait. All beings whose form is attained by dehumanization, not only stop acting like humans, but more and more closely resemble the image, we see, of the beast".

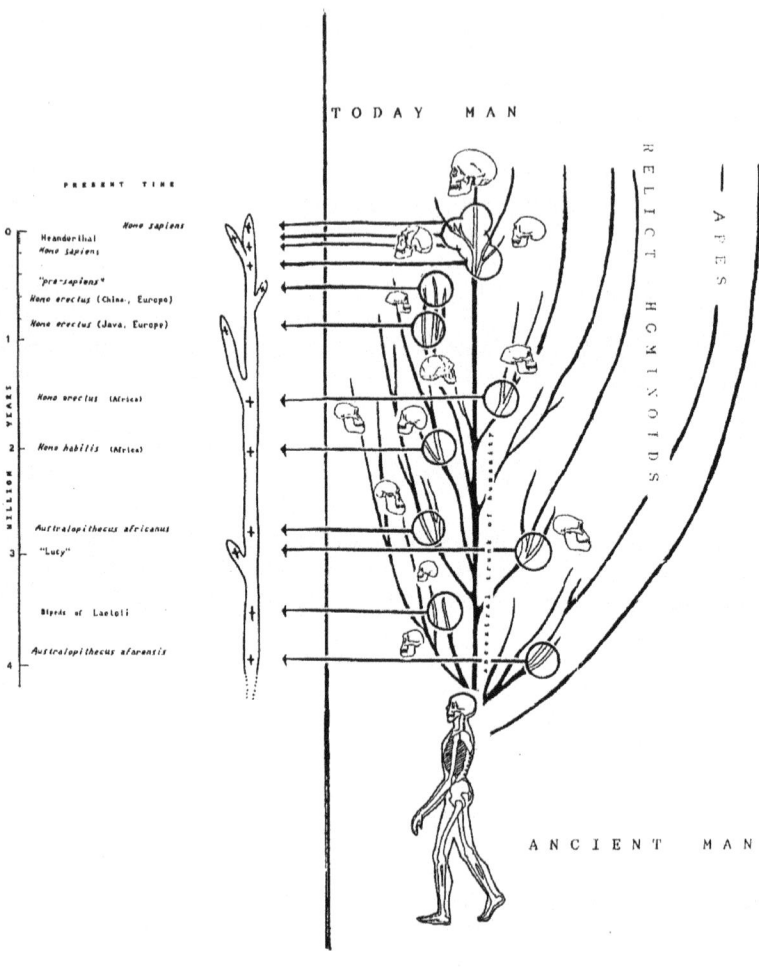

Fig. 4. Genealogic Tree of Man's evolution including the 'relict' hominoids. The author's personal impression is that the 'trunk' of the family tree consists of big headed *Homo* forms assuring specific continuity.

The Branches of the tree include many *dehumanised* creatures, known from the fossil record (left side), and/or continuing to exist as relict 'hominoids' surviving alongside the anthropoid apes. (right side)

What are the starting factors for dehumanization? First, there may be a breaking in cultural habits, then a change in eating habits, and finally a change of habitat. This is a concourse of circumstances that may very well happen after a big natural disaster. (Or in the case of humans a self induced catastrophe).

In his famous book on wild men and men of the woods, Ivan T. SANDERSON (1963), let us consider with new eyes, the condition of human groups that are rejected into mountains, forests or other inhospitable areas. In Norway, some adolescents who had grown up in humid valleys that were nearly always deprived of sun rays suffered from physical degenerations due to the lack of Vitamins E and D (produced by sun rays). Suffering from mental subnormality, they had grotesque hairs growing on their head and body, their jaws were large and projected forwards, the teeth were large and irregular. Rejected by the community such people lived in the mountains and succeeded in eking out an existence by hunting small animals by hand. They were eaten raw.

Conclusion.

Within the theoretical framework of initial bipedalism the relict hominoids are considered to have issued from the human line of ascent. They are not our 'ancestors', but the representatives of collateral lineages that have survived, concurrently with H.sapiens from prehistoric times.

The classical *yeti*, a big quadrupedal ape, perhaps shares the same dehumanized outline as the orang-utan and the fossil *Sivapithecus*.

The *kikomba* and other *kakunakari* of Africa seem to be effectively surviving australopithecines that have remained without notable change since the days of their appearance in the fossil record.

The giant hominids of *sasquatch* or *nyalmo* type are undoubtedly linked to the fossil *Gigantopithecus*, whose simian nature only exists in the imagination of the palaeontologists.

The *almasty, kaptar, bar-manu* etc., are contemporary neanderthalians who found refuge in the high mountain forests of Eurasia. In the expectation, perhaps of recovering their former range if the climatic conditions of the planet should worsen again.

In that case, man of *H sapiens* type would be confronted with survival problems of our own should we ever be forced to return to the caverns. When, as nowadays, the human species is in a phase of expansion, the hairy wildmen remain hidden. A good biological adaptation, a natural behaviour mixing mistrust, fear and also curiosity, and a location that remains very inaccessible to humans allows these creatures of the twilight to survive without too much constraint.

Here lies the zoological explanation to the problem of the hominoids we call 'relict'.

REFERENCES

BAMAIN, Jean-Jacques et al. (1993): *"Histoire d'Ancetres-La grande aventure de la prehistoire"*. Musees/Hommes, 3, 20, Paris.

COPPENS, Yves (1983): *"Le singe, L'Afrique et l'homme"*. Fayard, Paris.

FRECHKOP, Serge (1940): *"Considerations preliminaires sur l'evolution de la dentition des Primates"*. Bull. Mus. r. Hist. nat. Belg., 16. (11), 1-22, (fevrier), Bruxelles.

GRISON, Benoit (1990) *"Etat actuel de la question du yeti"*, Bipedia, 4 1-10, (mars), Nice.

HEUVELMANS, Bernard *"L'Homme des Cavernes a-t-il connu des Geants mesurant 3 a 4 metres?"* Science & Avenir, 63, 2-7 (mai), Paris.

HEUVELMANS, Bernard (1954): *"L'Homme doit-il etre considere comme le moins specialise des Mammifères?"* Sciences & Avenir, 85, 134-136, 139 (mars).

HEUVELMANS, Bernard (1958): *"Oui, l'homme-des-nieges existe"* Sciences & Avenir, 134, 174 (avril) Paris.

HEUVELMANS, Bernard *"L'enigme de l'homme congele"*, in HEUVELMANS & PORCHNEV: *"L'Homme de Neanderthal est toujours vivant"*, Plon, Paris.

HEUVELMANS, Bernard (1980): *"Les Betes humained d'afrique"*, Plon, Paris.

HEUVELMANS, Bernard (1986) *"Annonated checklist of apparently unknown animals with which cryptozoology is concerned"*, Cryptozoology, 5, 1-26, Tucson.

HEUVELMANS, Bernard (1993): *"Le dossier deos Hommes Sauvages et Velus d'Eurasie"*, 3" Milleniare, 28, 44-55, 66-67,; 29, 50-61. Paris.

HEUVELMANS, Bernard, (1993): *"Les creatures de l'Ombre"*, VSD-Nature, 5, 62-75, (decembre), Paris.

HUTCHINSON, Robert (1'991): *"Sur les traces du yeti"*, ed. Robert Laffont, Paris. ("In the tracks of the yeti", 1989).

KOFFMANN, Marie-Jeanne (1991): *"L'Almasty yeti du Caucase"*, Archeologia, 269, 24-43, (juin), Dijon.

KRANTZ, Grover S. (1986): *"A Species Named from Footprints"*, Northwest Anthropological Res., 19 (1), 93-99.

MAGRANER, Jordi (1991): *"Notes sur les Hominides reliques d'Asie Centrale"*, ed. Troglodytes, Valence.

RAYNAL, Michel (1989): *"L'Homme Sauvage dans les Pyrenees et la Survivance des Neanderthalians"*. Bipedia, 3, 1-16 (septembre), Nice.

RAYNAL, Michel (1993): *"Les Neanderthalians reliques des Pyrenees au Pakistan"*, Bipedia, 10, 14-24, (juin).

RAYNAL, Michel (1994), *"L'Homme Sauvage dans les Pyrenees"*, Cryptozoologia, 4, 1-8, (julliet), Bruxelles.

ROUMEGUERE-EBERHARDT, Jacqueline (1990): *"Les Hominides non idetifies des forets d'Afrique"*, Plon, Paris.

SANDERSON, Ivan (1963): *"Hommes-des-Neiges et Hommes-des-Bois, Les Primates ignores du monde"*, Plon, Paris, ('Abominable Snowman, Legend Comes to life', 1961).

de SARRE, Francois (1989): *"La Theorie de la Bipedie Initiale"*, 3" Millionaire, 12 a 17, Paris.

de SARRE, Francois (1991): *"Essai sur le statut phylegonique des Hominoides fossiles et recents: le point de vue de la theorie de la bipedie initiale"*. Bipedia, 7, 1-6, (septembre), Nice.

de SARRE, Francois (1992): *"Ueber die aquatile Lebensweise des Menschen in den fruben Zeiten seiner Entwicklung"*. Efoden News, 11, 13-15, (Oktober), Russelsheim.

SIEVEKING, Paul (1989): *"Of Human Tails"*, Fortean Times, 52, 52-54, (Summer), London.

VANDERMEERSCH, Bernard (1988): *"L'extinction des Neanderthaliens"*, in: Dossier histoire et archeologie. Archeologia, 124, Dijon.

WEIDENREICH, Franz (1945): *"Giant Early man from Java and South China"*, Anthropological Papers of the American Museum of Natural History., 40, 1-134, New York.

WESTENHOFFER, Max (1935): *"Das Problem der Menschwerdung"*, Nornan-Verlag, Berlin.

Francois de Sarre
President of the C.E.R.B.I (Nice)
Member of the Societas Europea Ichthyologorum (Frankfurt).

September 1994.

* * * * *

The Flying Snake of Namibia: An investigation.

by Richard Muirhead.

In south-eastern Namibia, in the vicinity of the settlement of Keetmanshoop and towns and farms within a radius of approximately eighty miles, there are reports of a flying, dragon-like snake. These reports have occurred since at least 1942 and probably much earlier. The presence of as yet unresearched cave paintings and reports from European Lutheran Missionaries are just two avenues for further investigation.

This article is a survey of all that is known by the author and his contacts covering a seven month period of research from April-October 1995. It is by no means exhaustive. The records of the diamond mining company which was active in the area, missionary and local newspaper reports were not consulted. (The latter are not available at the Colindale Newspaper Library in London). However, despite these drawbacks, fascinating information has come to light. What is particularly interesting are the paralells between this animal and the supposedly legendary dragon.

Unfortunately the account by a significant witness, Mr Michael Esterhuise in 1995 differs somewhat from the account of his experience published in 'These wonders to Behold' by Lawrence G.Green in 1959. The interview on the 1995 video appears to have been edited. It is also hard to know where reality ends and mythology begins (and vice versa). Readers should weigh up all the available evidence and decide the veracity of the reports for themselves and write to me with their comments!

It must be emphasised that this animal is probably not the same as the Oliatiaou of the Cameroons, which was reported in the early 1930's, or the pterodactyl like creatures sporadically reported from Zimbabwe ('Kongamato'), Kenya, Mexico and elsewhere.

> (NOTE: In 1995 Angus Whitty, a South African TV producer made a documentary about the putative creature for the NNTV network. He was kind enough to give me a copy of the film, which has never ben broadcast in the U.K. All the people in the paragraphs below appear in the NNTV Video)

The Flying snake has been seen in Kirris West, a farm nearly sixty miles east of Keetmanshoop. It has also been reported from the settlements of Berseba, Tses and Koes. (See Map on Page 113). It is amazing that anything can live in this, one of the most under-explored regions of the earth, because of its arid barren terrain. Namibia is the driest country in southern africa .... 'there is almost no rainfall'.(17).

Southern Namibia, close to the Kalahari Desert, is also arid, but there are thorn bushes and hills (krauses), where the snake is said to live in small caves.

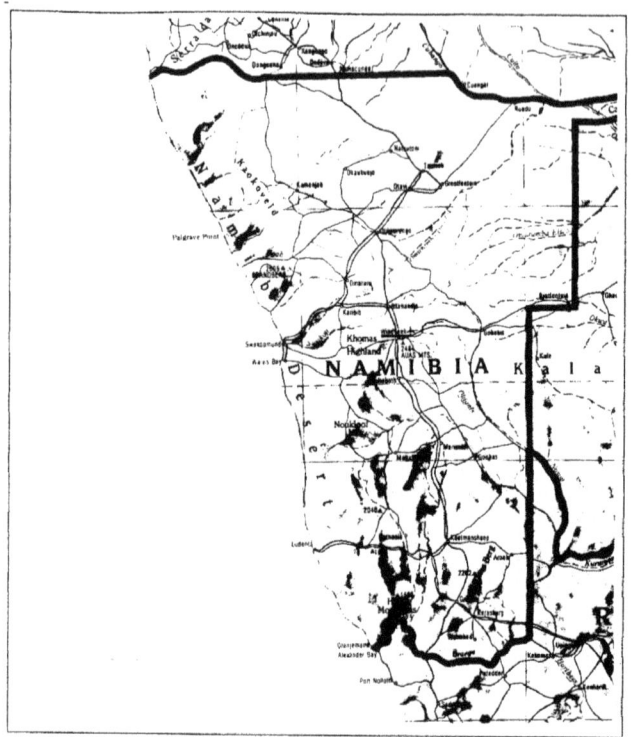

Namibia

It was from one of these caves that Michael Esterhuise was attacked in January 1942. Then sixteen, he was tending sheep, when...

"I heard a sound like wind blowing through a pipe, and suddenly the snake came flying through the air at me (....) it landed with a thud, and I threw myself out of its path. The snake skidded, throwing the gravel in all directions. Then it shot up in the air again, passing right over a small tree, and returned to a hill-top close by". [6]

This hill was three hundred feet high. This begs the question, 'how did it fly?' According to Roy Mackal's account, based on his summary of reports, the sound that the animal made was considerably louder than 'wind blowing through a pipe' it was a 'great roaring noise'.[8]

"In fact", (he writes), "it is hard to attribute such a disturbance even to a large, gliding creature, suggesting instead that some kind of wing action may have been involved".

The only flying reptiles in the world today are the flying lizards of the East Indies (*Draco volens*), and a mildly poisonous tree snake called *Chrysopelea*, which lives in South-East Asia.

(EDITORS NOTE: These reptiles should more properly be known as 'Gliding Reptiles' as they do not fly in the true sense of the word).

In the South African NNTV documentary [20] the artists impression, and Dickie Araun's description give the position of the wings at the side of the mouth. They are described as "*being like a bat's*".

How could such wings assist aerodynamically in flight? (Johanes Erman (a witness) indicated that the creature could hover about six feet above ground level. Could the 'wings' be a pronounced hood held up by extendible ribs?

Could gliding, or even flight, be enabled by

a) 'Hollowing' the body to promote air resistance?
b) Producing a gliding membrane out of the aforementioned hood? [13]

If the animal flies by making a concave shape out of its body, perhaps that surface would make it able to fly. If the animal has extensible ribs or rib like projections then it might look like the snake below.

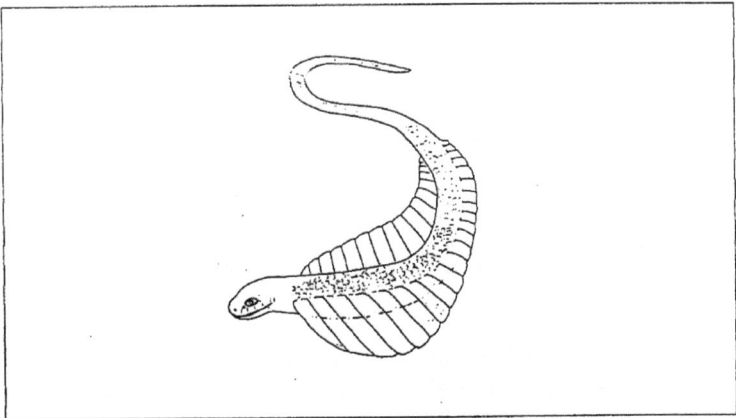

Pic: Courtesy Francois de Sarre. '*The Serpent aile*' de la Bollene-Vesubie. A Flying snake reported from Southern France during 1930/1.

The hypothetical 'wings' have to give it the ability to keep it airborne for more than a few feet. One report that may be true described the animal flying from cliff to cliff, and not just from the cliff to the ground. Michael Esterhuise, also described the animal flying UP the cliff.

The hood noted by some witnesses suggests that the animal may be a kind of cobra. Other reports, like that of Michael Esterhuise refer to an animal which projects itself forward with its tail, which it uses to slap from side to side.

Doctor Marjorie Courtenay-Latimer, of coelecanth fame, investigated the Kirris West sighting and spoke of it leaping and jumping, and a woman witness spoke of it 'taking off' from flat ground. How could a snake, which whether like a python or a cobra in shape, measures from between nine to fifteen feet in length and is up to a foot thick, fly without dragon or pterodactyl like wings? This mystery remains. I conclude this discussion with the viewpoint of a sceptic [12]

"I am familiar with all the large snake species in southern africa and would make the following observations: The python and the puff adder are too heavy to get airborne and (....) any attempts (by a python), to achieve the feat of flight would almost certainly result in its landing in a heavy heap on the floor directly below".

Another correspondent wrote:

"One of the most marked evolutionary trends in snakes has been towards limb loss, and there is no indication that they ever produced winged forms". [10]

The snake is also reported to have a light on its head. This may be a reflective scale, a white spot, a lighting up stone, a lamp or a mirror depending on the perspective of the viewer on on the mythological context that we view the ''creature' within. The light only shines at night, but the only direct witness that we have to the light did not see the snake itself. Michael Oarum says:

"I saw a light coming out of the mountains, and just wondered whether it could be an aeroplane or something like that, but after a couple of minutes I thought; 'but aeroplanes do have sounds. What kind of thing is this which doesn't make any sounds?' That's when I saw a big light in the clouds, and later on approaching slowly but surely 'till I was spotted (?) (word indistinct on original recording), in the light. It was kind of blinding me, so I turned my back on it. Shortly afterwards something told me 'its a big snake. Run for your life!' so I ran away" [13] [20]

This is not an eye-witness to a mystery snake, but there is, in the heat of the moment, a referral to a cultural belief linking a flying snake with a mysterious light. In the west we have a similar link between 'foo fighters', ball-lightning and mysterious lights in the sky. We shall return to the historical perceptive links between lights in the sky and 'dragons' later.

To corroborate, Dr Sigrid Schmidt, of the Namibian Scientific Society during the 1970's, says:

"Like the ghosts or UFO's in Europe, these snakes are seen by people who believe in them, and people who do not believe in them do not see them. A teacher once sarcastically told me:

If at night, people see the light of a motor-bike or a car where only one head-lamp is working, people say: 'Oh there is that snake again!'". (15).

Given the paucity of concrete evidence that there is a real light/snake link, it seems that the sightings of mysterious lights have been 'grafted' on to the stories of the mystery snake. Perhaps the snake, has become the explanation for some otherwise inexplicable lights. Our more technological society relates similar lights to hypothetical aerial craft manufactured by a civilisation with a superior level of technology.

Theoretically, however, there is still the zoological possibility that the snake may have developed a light on its head, possibly to attract its prey.

According to witness Andreis Fliemas, the snake can also change its colour. It is unclear whether this means that it has similar powers to a chamaeleon, or whether it merely has two colouration morphs. The colours were described by one witness as being brown and spotted, or yellow. Could it perhaps change colour according to the season? or from individual to individual? There is not enough evidence from either the video or the available literature to justify a conclusive statement on this.

The 'whiskers' and the horns and/or hood are another feature of its anatomy. In his *'Serpentum et Draconum historae libri duo'*, viper like snakes are featured with small horns above the eyes but without wings.

(EDITORS NOTE: Several species of viper boast what appear to be horns. These include the Desert Horned Viper *(Cerastes cerastes)*, and several other species across the globe.)

A mysterious snake which has been reported from both southern Africa (and, interestingly, the Carribean) is the crowing crested cobra. This creature, which has not to date been officially described *"allegedly bears a prominent, bright red crest, resembling a cocks comb, but projecting forwards rather than backwards"* (16). An excellent overview of this mysterious creature can be found in *'Extraordinary Animals Worldwide'* by Dr Karl P.N.Shuker (Robert Hale, London 1991).

The crowing crested cobra is also said to dwell in trees and amongst "krunns's" and large rocks.

Sir Laurens Van-der-Post, the legendary explorer wrote to me during 1995 on the subject of the Namibian Flying Snake:

"The snake takes many forms in the mythology of the black people of Africa, and this is not an unfamiliar one, although it has been difficult to pinpoint someone who has seen a flying snake. But I have, in Zimbabwe and Zambia, some fifty years ago, encountered Africans who swore that they saw frequently a crooning, crested cobra that could fly as well". (18)

We should now further consider the appendages. There is said to be a snake called the 'Noya a Thaba' living in the mountainous country to the west of the Kruger National Park which has

SIR LAURENS VAN DER POST CBE

4 August 1995

Dear Mr Muirhead,

I could write you a screed in answer to your letter; but just briefly: the snake takes many forms in the mythology of the black people of Africa and this, I think, is not an unfamiliar one although it has been difficult to pinpoint someone who has seen a flying snake. But I have in Zimbabwe and Zambia, some fifty years ago, encountered Africans who swore that they saw frequently a crooning, crested cobra that could fly as well. That is all I can tell you.

Yours sincerely,

Laurens van der Post.

Letter from Sir Laurens Van der Post to Richard Muirhead

If 'three feathers on top of its head'. [21]. It is said to live in certain 'kloofs' (caves) in the mountains. Some reports of the Namibian animal say that it has 'ears' instead of 'horns'. The 'ears' and the 'feathers' of the 'Noya a Thaba', could both be descriptions of the same organs. The habitats of the Namibian creature appears to be similar to that of the 'Noya a Thuba'; hillocks or 'kloofs'. The 'Noya a Thuba' has been described as living coiled up in the vicinity of trees, with its head resting on top of the coil.

As a result of its attack on Michael Esterhuise, the cave where it was supposed to live was dynamited, and a roar, like that of a lion, was heard. Earlier it had been reported to make a 'growling' noise. Roy Mackal's testimony suggested that it made a 'low moaning sound'. [8]. It is also reported to call like a sheep or a goat. (Investigators into mystery snakes elsewhere in the world have suggested that this would be the sound of the creature imitating its prey). In the snakes' lair at Kirris West the bones of lambs and bucks were found. By the time Dr. Courtenay-Latimer arrived on the scene, there was no sign of the snake but there was also no trace of the birds and small mammals which would normally be expected to inhabit the area.

The length of the Namibian snake is not particularly large and fits in easily with that of a cobra or a python. Witnesses find it hard to agree on the length of the snake. Sergeant Honeyborne, 'a very good narrator' [15], on checking all his evidence, back in 1942, suggested that the total length of the snake was about twenty five feet which would make it comparable in size to many large pythons, but not with the record holders. Other more conservative reports refer to an animal of between nine and fifteen feet in length. The largest venomous snake in the world is the Hamadryad or King Cobra (Naja hannah). One found in Malaya during 1937 measured 18 feet two inches, and was killed two years later as a precaution at the outbreak of war. At the time of its death it had gained an additional seven inches. [22].

Of its other features...

According to Michael Esterhuis, it left a foul smell, like tar or burnt brass. (This reference is a little confusing, as brass. being a metallic alloy cannot be burned!). According to its mythology, its smell alone can kill and attracts swarms of flies. Jonathan Downes and others have suggested that this could be the result of an excretion voided from its cloacal glands by a frightened snake. He also suggested the outside possibility that the fluid could be a lubrication medium produced to help a snake 'skid' along the ground after landing from a 'flight'. The video contained 1942 photographs of these 'skid' marks, which were curved like the tracks of some vehicle. Dr, Courtenay-Latimer, at least, was convinced that these were made by some large reptile. She, however, suggested that they could be as the result of abnormal behaviour by a wounded python, which had possibly strayed from a tributary of the Orange River.

Here we should consider the historical aspects of the matter. the earliest evidence of a similar animal is from Pliny who said that the Basilisks of Cyrene have a white spot on their head representing diadems.

Apart from the legends of the Chinese dragons and similar creatures, I have noted several significant paralells between the Namibian snake and and the dragon entry in T.H.White's 'The Book of Beasts'. [19].

- The dragon comes out of a cave.
- The dragon is carried into the sky with the air around it becoming ardent (my italics).
- It has a crest.
- Its strength is in its tail.
- It uses its tail to inflict injuries rather than using poison.

All these characteristics are to be found in the Michael Esterhuise case.

I am especially interested in the link between the 'ardent air' and the smell of tar, often linked with reports of the beast.

Dragons are described as killing their prey in the manner of a constricting snake such as a python. The bestiary also says of dragons that:

"..they are bred in Ethiopia and India in places where there is perpetual heat". [19]. *The reference to 'Ethiopia' probably refers to the whole of 'black' africa, as opposed to 'Arab' or Egyptian Africa).*

Even earlier than the above references is the small repository of dinosaur-like animals described in the pre-Christian saga of Beowulf. [5]. Beowulf, who was born c. 495 AD, died c. 583 AD, died whilst in combat with a Widfloga, a giant Pteranodon-like creature which died at the same time. [19]. A fuller account of this episode is given in the Creation Science Movement pamphlet number 280, entitled 'Anglo-Saxon Dinosaur'. The Creation Science Movement can be contacted at 50 Brecon Avenue, Portsmouth, Hampshire, U.K. PO6 2AW. This seems dis-similar to the Namibian snake, but is a possible lead for future research.

By the period between 1487 and 1499 the Portugese had sailed down the Atlantic coast of Africa, and past the Cape of Good Hope. Is it conceivable that these early explorers brought accounts of giant flying snakes back to Europe where they became entangled with both contemporary and historical accounts of huge reptiles. Which came first, the chicken or the egg? The European or the African dragon stories?

In c.1591 a Portugese adventurer, Odvardo Lopez explored the Congo region. Notes of his explorations included dragons of azure blue and green [7], and covered in scales. These dragons had an unfamilar feature - only two feet.

Here it should be noted that in an obscure passage on the NNTV video, an eye-witness speaking to researcher Marcus Oarum drew a link between his (Oarum's), bright blue (azure?) clothes and the brightly coloured snake. Here, perhaps we should also point out that certain types of lizard and salamander, not to mention cryptids like the European Tatzelwurm and several mythological dragons have only a pair of front feet.

Interestingly the Congo is now home to a contemporary 'dragon', the hypothetical prehistoric survivor, Mokele Mbembe. A fuller treatment of this case can be found in *'Les Dernier Dragons d'Afrique'* by Bernard Heuvelmans [7].

-Earlier that century, in 1560, and again in 1569, 'Flying Dragons' were also seen. These were possibly meteors in the lower atmosphere; fiery lances and pillars of flame. They were called Draco Vulcus! The 'Kalender of Sheepeherds' (sic) published in London in about 1560 mentions them. Here we have an interesting parallel with the situation in Namibia where there is a clear link between mysterious lights in the sky and dragons/flying snakes.

In 1614, a dragon roamed St Leonards' Forest in Sussex, two miles from Horsham. The following passage is from the 'History and Antiquities of Horsham' by Dorothea Hurst (See reference 2):

"There is always in his tracks or path left a gluytinous and slimie matter, (or by a small similitude we may perceive in a snake), which is very corrupt and offensive to the scent, in so much that they perceive the air to be putrefied with all which must need to be very dangerous".

Later, in 1667, Luys del Marmol Carvajal cited in Dr Owen's essay, 'Towards a Natural History of Serpents' (1742), reported an infinite number of winged dragons, supposed to be incredibly poisonous, in the Atlantic or Atlas caves and mountains in Africa. The same work by Dr. Owens, (which is admittedly a rather questionable source, repeating earlier sources in Natural History), describes the Horn Snake. This reptile reputedly struck at its enemies, different snakes, with its tail. He also reported that some are covered with sharp scales which make a 'bright appearance in some positions'[1].

In 1840 a young man was walking home in darkness near Sesterbach in the Eifel area of Germany, when a 'fiery dragon' hit him. He died soon after from the burns[9].

In 1930 or 1931, a winged snake-like animal frightened the wife of a road worker in a hut in a highland forest at La Bollene-Vesubie near Nice in southern France[14]. It flew down from the branches of a tree outside the window sill! The animal reminded Francois de Sarre of the tarasque of the lower Rhone valley, and also of Draco volens, in a similar way as did the Namibian snake to me. In 1984 the son of the original witness found a pot in Cervasea, Italy with a flying serpent on it![14].

There is, by the way, a motif in a church in Hadfield in Derbyshire which apparently resembles the Tarasque. It has its wings folded from the back to the sides of its body (like a hanging bat?) in a similar stance to the creature at Tarasque de Noves in the Avignon museum. There is a Derbyshire folk story connected with the Hadfield Church motif, but I shall leave that to a future researcher.[5]

Finally, I would like to mention two episodes from 1975.

According to a Swedish Museum official a pterodactyl like animal was seen in a swampy area of Namibia during late 1975. Also in 1975, (according to the Swedish newspaper 'ABC' - July 17th 1975), a curious 'meterological phenomenon' appeared over Gerona, Spain. A glowing form like a monstrous head with a dragon's tail was seen at 10.15 pm[9].

I have deliberately avoided mentioning the flying snakes of Egypt or Arabia which were mentioned by Herodotus and others. I have also avoided mentioning the Welsh flying snakes from the turn of the century, and the mystery flying snakes of eighteenth century London, mentioned in Michael Goss's *"Flying Snakes and Winged Serpents"* (Unknown Magazine 1986), because these species are so much smaller than the animal reported in Namibia.

(EDITORS NOTE: For discussions on the smaller flying snakes, and for further discussion of 'Fire Drakes', I would refer you to Dr Shuker's article elsewhere in this yearbook. For further discussion on the St Leonards Forest "dragon", see Roy Kerridge's article, also in this yearbook).

As far as the dedicated mythologists are concerned, the question is this:

"Should we ask what the snake 'means', what it represents, or what it symbolises, rather than whether or not it is 'real'?" [11]

In Africa snakes can signify an association with a human desire. There is also, as far as the Namibian snake is concerned a number of other links. The Nama see a link with water, a danger to man and a link with the darker areas of the human psyche. It appears to be similar to the Australian 'Rainbow Serpent', with a definite connection with water in both cases. The Namibian snake is one of the foremost topics of Namibian Folk Belief, and legend. There is an equivalent in the desert areas of the country, called the 'big snake', and there is an equivalent, said to live in the rivers. These are the barest bones of the mythology of mystery snakes, and I would stress that it would require a completely separate essay to attempt to do justice to the matter.

In conclusion. What is to be made of the mystery flying snake of southern Namibia? Even if the sighting near Kirris West is the only well documented case, and a large percentage of the reports are influenced by the above mentioned mythological thinking, it would still be significant. Dr Courtenay-Latimer tentatively endorsed the mystery snake, and its existence is not confined to the local Nama. Moreover, its landing tracks were photographed at Kirris West. Similar tracks were also reported to have been photographed after a flying snake is reported to have killed, or at least knocked down a diamond prospector after flinging itself off a sand dune during the 1940's. The photograph, if it exists, must be in the archives of some newspaper, but to date I have been unsuccessful in locating it. [11]

Writing in 1959, in *'More Animal Legends'* (Frederick Muller, London, 1959, pp. 116-131), collates a number of eye-witness reports, including his own observations of a snake's ability to rear and jump. We must also bear in mind the interesting parallels between this snake and dragon legends.

The discovery of a 'flesh and blood' flying snake, or even a fossil of such a creature, would radically call into question, the presently accepted evolutionary model for snakes. It would be marvelous for zoology if the century which began with the discovery of one remarkable African creature, (The Okapi in the Congo in 1901), could end with the discovery of an equally remarkable African animal, the giant Flying Snake of Namibia!

The author would welcome feedback c/o the Centre for Fortean Zoology. He would like to thank everyone who helped with his research, and would like to wish Marcus Oarum luck with his dream of obtaining concrete evidence of the Namibian animal.

REFERENCES.

1. OWEN, Dr. Charles, 'An essay towards a natural history of serpents'. p.4. (Pitchfork Press 1965 reprint of 1742 original).

2. ANON. 'The story of the forest', (Ed. Unknown, St. Saviour's, Culgate; publ. Unknown). Quoting HUNT, Dorothea, 'History and Antiquities of Horsham'. Other details unknown.

3. ASHTON, John. 'Curious creatures in Zoology'. (New York; Cassel Publishing Company, p.319).

4. BEAZLEY, Mitchell. 'Mitchell Beazley's: the world atlas of exploration'. (London; Mitchell Beazley Publishing Ltd, 1975, p.74).

5. COOPER, Bill. 'Anglo Saxon Dinosaurs'. (Creation Science Movement, Pamphlet No. 280. First Edition. Portsmouth; Creation Science Movement, 1992).

6. GREEN, Lawrence G. 'These Wonders to Behold'. (First Ed. Capetown; Howard Timmins, 1959. pp.186-8).

7. HEUVELMANS, B. 'Les Derniers Dragons D'Afrique. (Libre Plon, Paris, 1978. pp.408-10).

8. MACKAL, Roy P. 'Searching for Hidden Animals'. (London; Cardigan Books, 1983 pp. 51-54).

9. MAGIN, Ulrich. 'European Dragons; The Tatzelwurm'. (Pursuit. Number 73, Vol. 19 No. 1. First Quarter 1986, pp. 19-20).

10. McCARTHY, Colin. Personal Correspondence, July 1995.

11. NICHAUS, Isak. Personal correspondence late summer 1995.

12. O'SHEA, Mark. Personal Correspondence Aug 1995.

13. PARSONS Sally. 'Position Paper on Namibian Flying Snake'. (Unpubl. Summer 1995).

14. DE SARRE, Francois. 'Are there still Dragons in Southern France?' (INFO JOURNAL. Autumn 1994. Arlington; 1994 pp.44-5) and "Von Drachen und Geflugelten Schlangen in den Franzosischen Stuttgart; Seealpen". (Proteg News, 3; 1995 pp.29-36).

15. SCHMIDT, Dr. Sigrid. Personal Correspondence September 1995.

16. SHUKER, Dr. Karl, P.N. *'Extraordinary Animals Worldwide'*. (Hale, London 1991; p.32).

17. SPARK, David. *'Power to the People'*. (Geographical Magazine, June 1995, p. 18-21).

18. VAN DER POST, Sir Laurens. Personal Correspondence August 1995.

19. WHITE, T.H. *'The Book of Beasts'*. (STROUD; Alan Sutton, 1992. p. 166-7).

20. WHITTY, Angus (dir). *'In Search of the Giant Flying Snake of Namibia'*. (Johannesburg; NNTV. 1995).

21. WOLHUTER, Harry. *'Memories of a Game Ranger'*. (2nd Ed. Johannesburg; Wildlife Protection Society of South Africa, 1949. pp. 242-3).

22. WOOD, G. *'The Guinness book of Animal Facts and Feats'*. (1st Ed. Enfield; Guinness Superlatives Ltd. 1972 p.190).

* * * * *

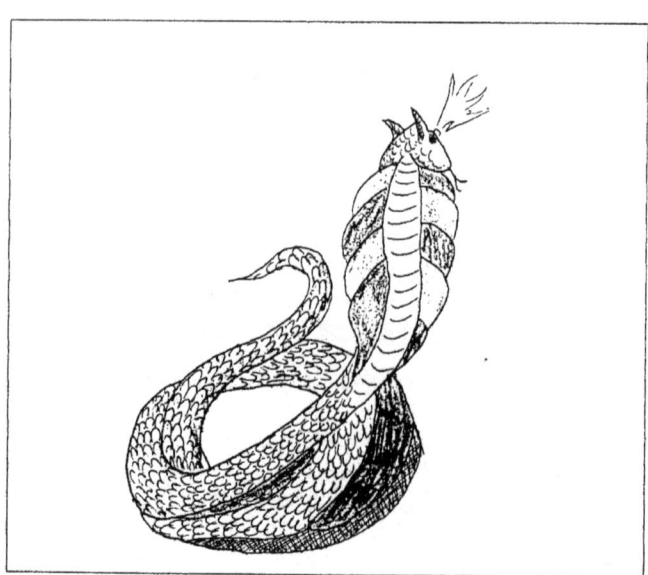

THE FLYING SNAKE OF NAMIBIA AS DEPICTED ON THE NNTV VIDEO

BOOK REVIEWS

(EDITORS NOTE: We receive a lot of books for review. Unfortunately we only have a maximum of two pages per issue that we can devote to book reviews and so various books of interest have perforce been left out. I am pleased to be able to include them in this section together with some reviews (in English) of German Language publications sent in by Hermann Reichenbach. The CFZ is planning to issue a free supplement to 'Animals & Men' sometime in the summer of 1996. This will feature reviews and details of a large number of books about both natural history and forteana).

'TROLLE, YETIS, TATZELWURMER - RATSELHAFTE ERSCHEINURGEN IN METTELEUROPA.'
('Trolls, yeti's, winged dragons: Mysteries of Central Europe')

by Ulrich Magin

(C.H.Beck pub. 1993. 184pp. 18 illustrations. pb. DM16.80).

Yetis are seldom thought of as a Central European phenomenon, but then few in Europe would admit to haviung seen trolls or winged dragons either. Magin - or his publisher - has chosen to title his newest Fortean book using three well-known 'representatives' as 'sales vehicles', if only because Charles Fort himself is an unknown commodity in the German Speaking community.

Perhaps a third of this pocket-book could be called cryptozoological (sea-serpents and dragons, unknown felids and canids; no yetis); the rest is devoted to a range of Fortean chapters on items such as goblins and giants, raining fish and globular lightning . Everything is well documented; nothing is new. Perhaps Magin's greatest service with this book was to have coaxed one of Germany's most prestigious publishers into issuing a book on a subject most would not touch with lead!

H.Reichenbach.

'WISSENSCHAFT UND FABELWESEN - EIN KRITISCHER VERSUCH UBER CONRAD GESSNER UND ULISSE ALDROVANDI'
('Science and Fabulous Creatures: a critical study of Conrad Gessner and Ulisse Aldrovandi')
by Christa Riedl-Dorn. (Bohlau, 1989, 183 pp,

18 illustrations. pb. DM68.00).

Zoology has many fathers; the 16th Century naturalists Gessner and Aldrovani were indisputably amongst the most germinating. Gessner, 'the German Plinius' (when Switzerland was still a part of Germany), and the Italian Aldrovani, founder of a private natural history museum in Bologna, widely acclaimed as a renaissance 'Weltwunder', devoted space and time to the natural history of mythical animals.

Mrs Reidl-Dorn, now head of the archives at the national Austrian Natural History Museum, has produced a fine, scholarly thesis on Renaissance attitudes towards sea-serpents, dragons, wildmen and the like, and their place in Zoology

Volume six of 'Perspektiven der Wissenschaftsgeschichte' ('Perspectives on the History of Science') includes an essay on 'magic as science'.

A comprehensive bibliography and notes, and a good index (oddly enough, rather unusual in German non-fiction; totally lacking in Magin's book), round out a good background history for any Fortean Enthusiast.

(Series editor Helmut Grossing).

H.Reichenbach.

'The Sceptical Occultist

by Terry White

(Arrow pb 314pp £5.99).

A mildly interesting, and relatively scholarly description of a number of para-fortean occurences which have been more entertainingly described elsewhere. The main bulk of the subject matter is split between ESP and quasi religous apparitions, and is therefore of limited interest to the fortean zoologist. JD

'The Druids'

by Peter Beresford-Ellis

(Constable pb 304pp large format £9.95).

Excellent overview of celtic culture which overthrows many of the cultural preconceptions which most of us have been taught about the druids. What is even more impressive is that Beresford-Ellis has managed to write such an informative and academic tome, which is mostly sympathetic towards its subject matter without lapsing into the new-age drivel whicso often surrounds any discussion of things druidic. Even the final chapter about the recent revival of druidism is sensibly written, and doesn't mention 'Ozric Tentacles' once! Excellent and highly recommended. JD

'The Lost Houses of Eggesford'

by Matthew Axe, Lesley Chapman and Sharon Miller

(Forest Enterprises 80pp pb £3.99).

This is a wonderful little book which details a painstaking piece of historical detective work which eventually unearthed the truth behind a puzzling, and largely unknown mystery of English History.

Cryptozoologists could learn a lot from the methodology of these three investigators, and there is even a brief mention of an otherwise undocumented Alien Big Cat sighting. I cannot recommend this book highly enough.

For details telephone 01769 580250.

* * * * *

Loch Ness - a cauldron of definites and possibilities

by Neil Arnold

Mysteries are the magical ingredient that make this world interesting and save it from being engulfed by banality. However, frustration comes as part of the package. Although a mystery lasts longer than any lifetime I find it hard to accept that I will go to my grave not knowing the answer but I still feel that all mysteries should remain just that way. Of course, times come when we feel that we are a little nearer but in reality we never know where we are. The fate of the mystery no longer lies in our hands, for those mysteries seem to be way beyond our capabilities. I have come to this conclusion simply because every mystery speaks for itself and nothing else is going to change that. This is not a negative view on my part and I love being involved with investigations towards mysteries like the Bermuda Triangle, UFO's, ghosts, and most importantly, monsters.

Mysteries are simply too great to be solved. Man is destroying many things but there must be some wonderful segment of this existence that is not battered. However, due to all the fantastic things, I simply wish that human life could extend to the point where we will never miss any sort of event. I don't want to see any sea serpent dragged up in a net and cut to pieces by the same brutes who slaughter whales. I desperately want the mystery and the magic to baffle us all forever...but I just hope that the afterlife allows us to continue with our work and our admiration.

The Loch Ness phenomenon is a lasting mystery. We know that something is down there and that it is not part of some mass hallucination. Neither is it any sort of natural phenomenon. What lurks in those black waters is a monster. The enchanting, yet eerie habitat has caused people to throw up a number of possibilities, but now, at last, there is a feeling that people are accepting the existence of a large creature. I strongly believe that a number of unknown and monstrous creatures inhabit the loch, and I also feel, that although I believe there is a 'living dinosaur', there may also be a number of outsized and fantastic recognised creatures. I am not putting down the numerous reports as natural phenomena because I believe in the monster. I feel, however, that if there is, for example a plesiosaur living in the loch, there are other creatures as well, and it is extremely obstinate to rule out the existence of other creatures, for example, giant fish. People have been baffled by the various different types of report - 'the upturned boat', 'a long neck', and 'an undulating body'. These differing reports do not seem to point to one creature alone, and I would suggest that more than one species is responsible for the sightings. I'm sick of those who condemn 'Nessie' as a mere myth, but one has to accept that there are 'recognised' creatures out there whose appearance can very easily be misinterpreted as that of the 'monster'.

An assortment of creatures could be mistaken for the monster, but only if they are huge themselves. I feel that if anyone sees a splash, a small 'hump' or even a flipper they should not necessarily class this as a monster sighting. Unfortunately, science as a whole, is reluctant even to admit the existence of out sized individuals of 'known' species let alone unknown ones, so for anyone who is blindfolded; a 30 foot lake creature can be an otter, and a huge flying beast with leathery wings is a grebe! The time may come when the unusual is more easily accepted, but many will continue to feel that the accepted rules of science are accurate. This is unlikely, when you consider the number of encounters with odd and freak creatures as well as with oversized, yet recognised beasts.

In the oceans the depths are boiling, in turn causing all manner of monstrous life forms. These are too much for some people to accept. We often speak of man's progress, and how he will conquer the skies, but compared to the diversity of nature and what has gone before our achievements are insignificant. Our knowledge of the planet is minimal but we feel that if we destroy it and become obstinate towards it we shall conquer it. This is a poor view, and despite our rigorous and rapid destruction of the planet I feel sure that nature will prevail. It is nature, after all that makes the rules.

We are all aware that the oceans are truly magnificent, yet it seems as though many sceptics are unaware of the depths of the lakes of the world. People are not following a trend by reporting monsters. The hoaxers and bandwagon jumpers are there, and there are still too many people who expect clear film footage of every sighting. I certainly believe that many lakes are the homes of large, unknown creatures. I feel that the mystery of lake-monsters has become a little stereotyped, but then if some of these creatures are plesiosaurs, then we must accept this. People find it hard to accept the possibility of lake 'monsters' despite the fact that the lake is ancient, the waters black, and the depths are beyond our probing. There have been 'deepscans' and searches, but we must treat these creatures with respect and not dismiss their existence just because they fail to show up every time someone looks for them. This elusiveness is a natural part of the mystery and people should no longer criticise an enigma simply because the animals are so elusive.

The possibilities of 'known' animals which could be responsible for some of the Loch Ness sightings are not endless. 'Nessie' is not a wale or a sturgeon. The sturgeon seems to be a popular solution. This fish, which does grow to a very large size, could only be the source of the 'upturned boat' reports. Sturgeons have a series of triangular fins along their backs, and it is impossible to see how these could be confused with the horse like mane sometimes reported running down the neck of 'Nessie' and other lake monsters. Although they are of monstrous size they are not real contenders for the 'Nessie' crown.

My personal opinion is that giant eels are responsible for some of the sightings. Scientists claim that giant eels cannot exist in fresh water. My faith in science is not strong, however and its attitude towards so many unknown beasts has left me 'following my own nose' in such matters rather than being one of the people who cannot see past the end of it.

In his book *The Monsters of Loch Ness*, Roy Mackal lists various contenders for the truth behind the 'Nessie' enigma. I agree with his view that one cannot dismiss anything purely because some people consider it to be unrealistic. This is why we cannot rule out certain candidates for the identity of 'Nessie', purely because the concept may seem ridiculous to some people.

So, let us 'brush away' the rules of size for the moment and pretend that any creature can grow to a certain 'over-size'. The possibilities seem endless. It is certainly possible for eels, sea cows, sturgeons, whales, sharks, catfish and rays attain monstrous proportions. These days, however it is certainly uncommon to find a 30 foot eel. Although we are looking at a freshwater habitat, there are sharks and rays, and indeed some freshwater catfish, that would be perceived as 'monster' rather than fish by someone with limited knowlege of the field. I like the eel solution because it fits the available evidence better than the sturgeon does.

I feel that many of the suggested solutions can be ruled out. The 'upturned boat' sightings cannot be explained by an eel, but the stereotypical description of the long neck could be an eel. Unfortunately 'Nessie' reports have included such features as horns, huge flippers, horse like heads, manes, humps and leathery skin and these have been responsible for all sorts of bizarre suggestions as to its identity. Indeed if we accept all of these descriptions it is impossible to guess quite what might be lurking in the depths of the loch. Although there are believers in the plesiosaur theory, I am one myself, there isn't really even a solid base to work from and if one is willing to accept the idea of a dinosaur type creature then many new avenues are opened. We should accept that a dinosaur type creature could have survived, and thankfully there are some people prepared to accept such alternative suggestions. I suggest the 'eel' theory for those who could also believe in a monstrous, but more 'natural' creature.

A decade or so ago, off the coast of Dover, a group of divers were asked to swim down to retrieve some pumps from a pump-house that had become flooded. These divers searched the gloomy area but one diver encountered more than the obvious difficulties inherent in the task. He was confronted by a number of large eels, one of whom had a head the size of a chair. The diver returned to the surface without the pumps and vowed never to return. Other divers were, apparently, due to continue the search for the pumps, but I don't know what happened to them.

A similar tale comes from the river Clyde, where a group of sailors were fishing from a pier. One of the sailors hooked something so large, that it scared the wits out of him and he cut the line. Both these events, however were in sea water, rather than in the fresh water of Loch Ness.

Although I have followed the mystery, read the statistics and the figures and been bombarded with the views, the tales, and the supposedly accurate rules, I find myself at my conclusion because the eel seems the only logical explanation. One may see the 'eel' hypothesis is frail because Conger eels are not known to survive in fresh water, but it is no more unlikely than the views of the sceptics, or those, such as myself, who tend to support the plesiosaur theory. Giant eels do exist around the world, and probably also live in Loch Ness. An incredibly thick eel coming to the surface and then diving could possibly create a curve, whilst the 'humps' fit in well with its undulating body, and eels, like 'Nessie' have been known to take to the land!

My theories are just another ingredient to the stew which has boiled away for hundreds of years. Thousands of questions have been asked and not all have been answered. No theory has been proven, and I sincerely hope that it never will be. If the 'monster' were something that could be seen in an aquarium it would hardly be the great discovery we are hoping for. Unfortunately, as well, the policy of 'agreeing to disagree' does not help us in our quest for the solution to this riddle.

I am sure that many other theories will emerge, only to be pushed down by sceptical views. People will, however, continue to film the monster and the interest
will never die down. For some 'Nessie' is a tourist attraction, for others it is a puzzling, yet exciting phenomenon, and for others the mystery does not even exist. Even accepting the existence of monster sturgeons in some of our northern lakes is too crazy a hypothesis for some.

I have always tended to follow my own instincts rather than to base my ideas solely on the countless reports in books, the various details, and all the hokum surrounding the mystery. The experts are the best bet for that. It is my opinion, however that 'Nessie' could be a giant eel.

The debate has emerged within the pages of 'Animals & Men' and although it is frustrating, it is very exciting. Large creatures do swim the oceans and lakes, and some seem extremely out of place, yet they survive. We could venture down all sorts of avenues in this mystery and come up with all sorts of different conclusions. This report, however is simply based on a view I have always held.

In A&M5 Nick Morgan mentioned the "whales" in the loch as well as various other reports, which enter the realms of folklore with his mention of tales aimed to frighten children away from the waters. Just by seeing this I am fully aware of the confusion that the mystery builds, and that everyone has their own opinion. Unfortunately when considering the sturgeon along side the 'Nessie' characteristics I see a very frail comparison.

There are giants on the land and in the sea and mystery surrounds them all. I'm sure that my beliefs will be questioned and I fully realise that even a giant eel may not provide all the answers. It is, however, pretty close.

Loch Ness is a mere puddle compared with the mighty oceans, yet man cannot even satisfactorily explore the 'puddle'. Whatever creatures lurk there, whatever possibilities are thrown out, and whatever beliefs we have in our minds, we are never really here or there. I'm sure that these 'monsters' will outlive us all. Thank goodNESS! Wonders will never cease.

*****

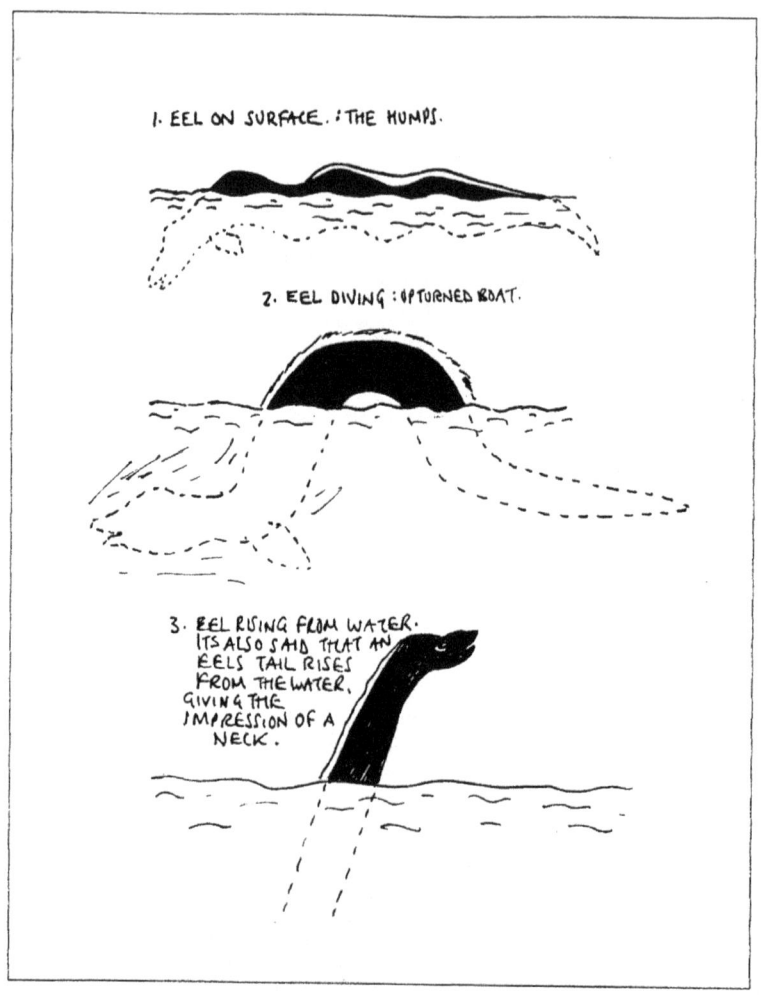

Sketches of how an Eel could be the 'monster'.
Roy Mackal describes the eel/'Nessie' comparison in his "The Monsters of Loch Ness".

Notes on the Whaling Industry at Peterhead.

by Tom Anderson.

(EDITORS NOTE: The history of British cetology is a chequered one. Perforce, the best source of historical records of cetacea from British waters is the whaling stations themselves. The history of the individual fisheries is equally important. As there is a certain cross over between cetology and cryptozoology we have been collecting data on the British whaling industry. We asked our Scottish correspondent for some information on his local whaling station. As usual he did us proud!)

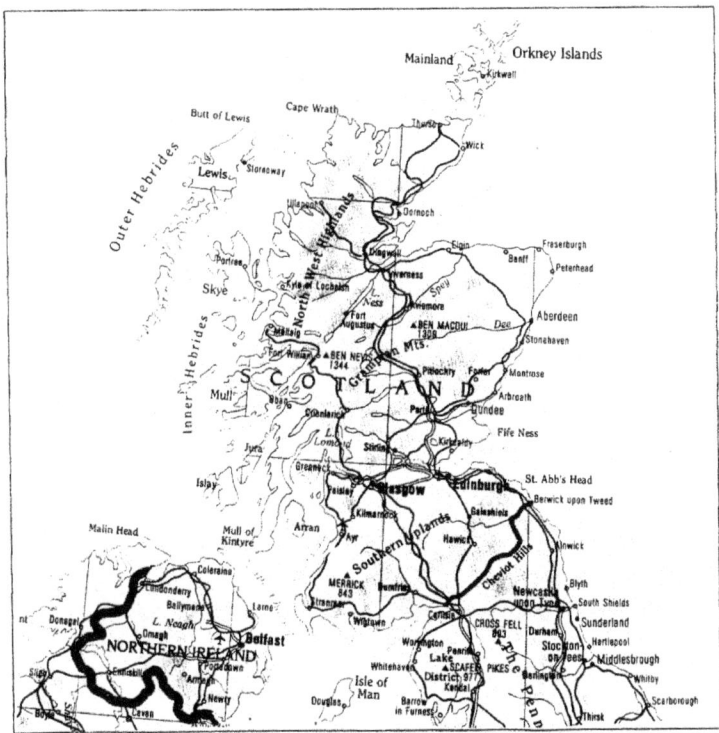

Scotland, showing Peterhead

- The first written record of British Involvement in whaling was off the Norwegian coast in the 9th century. The Basques had a thriving industry by the 13th. After depleting local waters they were catching whales off Newfoundland in the 16th Century. In 1613 the Muscovy Company trading from London was whaling off Spitzbergen up until 1623 when Dutch competition drove them out. By 1720 the area was fished out and attention turned to West Greenland and the David straits.

- Americans entered the industry at this time, sailing from Nantucket and New Bedford etc., but specialised in tropical Sperm Whales, rather than Humpbacks, Narwhals and Greenland Right Whales caught by Europeans.

- Hull was the principal British Whaling Port 'till the 1840's, then Peterhead until the 1870's and then Dundee until the first world war. Having grown from two ships in 1673 to over 250 in the 1780's, the UK fleet declined from the 1857 level of 60 until its demise early in the 20th Century.

- A typical oak-built whaler out of Peterhead was the 'Eclipse' of Captain David Gray. She was 149 feet long with a 29 foot beam and a hold depth of 16 feet. Three masted with a 75 h.p steam engine, she weighed 436 tons gross, and 295 tons net.

- The crew of between 24 and 42 from the mainland and the Shetland Islands sailed in February, returning either when full (sometimes as early as July), or at the season's end in October/November. At the end of the voyage, the crew received a certificate of discharge (Form DIS.1) stating their job title and period of employment along with the tonnage of the vessel and its master's name.

- Each vessel had eight small boats to close in on the whale in pairs. As each could only stow 600 fathoms of 2.75 inch line in the bows the second would harpoon the whale when it re-surfaced after half an hour. The other boats would then throw their lances before towing the dead whale tail-first to the mother ship.

19th Century print 'celebrating' the advent of the steam harpoon.

- David Gray, mentioned earlier, in forty-three years, took one hundred and ninety-eight whales, (mostly Bowheads (*Balaena mysticetus*) and one hundred and sixty eight thousand, t nine hundred and fifty six seals, mostly Harps *(Pagophilus groenlandicus)*.

- The whales yielded over a hundred tons of baleen reaching from £400 to £2,000 a ton and a corresponding quantity of oil.

- The earlier harpoons were copies of Innuit types, but were changed to swivel heads to improve retention. The harpoon gun was invented in 1731 and had a range of one hundred feet.

- Mortality was high in the early years. In 1830, out of ninety-one ships through the Davis Strait, nineteen were lost. The following year another six ships were lost and over six-hundred men were trapped on the ice. One hundred and thirty-five of them died of exposure and scurvy.

- Profits were variable. The 'Robert' in 1788, caught only three whales in four seasons.

- At the turn of the century a few of the crew stayed on the ice to organise the Innuit in hunting bears, walrus and narwhal until the ships returned in the spring.

- An average whale took four hours to 'flense'.

- One hundred and ninety-two tons of oil equates to thirteen thousand Harp seal pups taken by one ship in 1870.

- No income from Arctic products appear in the Financial statements of the Peterhead Harbour Trustees after March 1907.

- By far the largest number of whales caught were Bowheads, known as Greenland Right Whales, *(Balaena mysticetus)*, because, they were the 'right' whale to catch.

- The remainder were Humpbacks, Beluga, Rorqual, or Narwhal. As the industry went into decline anything was fair game and with the help of the natives Seals, Fox, Walrus and anything likely to help offset the costs of the voyage was taken.

- The devastation of the Harp Seals in particular was colossal and only the advent of mineral and vegetable oils alleviated their predicament.

(There then follows a very funny paragraph highly offensive to anyone from Norway and/or called Sven that I regretfully decided not to print it).

Sources of information on Peterhead Fishing and Whaling.

BUCHAN, A.R., 'Port of Peterhead'. (1980) - detailed accounts with selected statistics, of fishing and whaling.

BUCHAN, A.R., 'Fishing out of Peterhead' (1986) - summary account.

BUCHAN, A.R., 'The Peterhead Whaling Trade'. (1993) - modern detailed account of Peterhead whaling.

CREDLAND, A.G., 'Whales and Whaling' (1982) - summary account of history of British Whaling.

LUBBOCK, B 'The Arctic Whalers' (1937). -compendium of information on whalers - puts Peterhead whaling into national context.

SUMMERS, D.W., 'Fishing off the knuckle' (1988). - most accurate account of the history of the fishing settlements in the Peterhead area.

SUTHERLAND, G., 'The Whaling Years: Peterhead (1788-1893).' (1993). - latest account of Peterhead whaling.

ANON. 'The Ports of Peterhead, Handbook 1989-90'. (1989). - gives selection of up-to-date statistics on Peterhead harbour.

FINDLAY, J.T, 'A History of Peterhead'. (1933) - good chapters on herring fishing and whaling.

* * * * *

The Marauder of Talog.

by Ian Hawthorne.

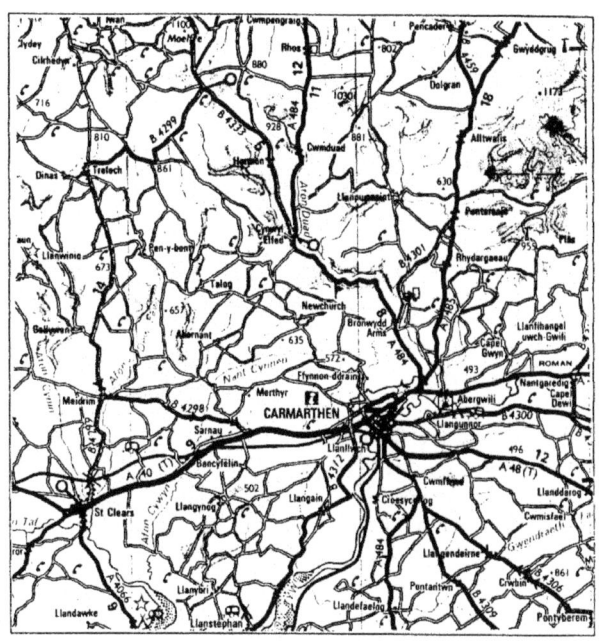

Date: June 15-July 16, 1987
Location: Talog, North-West of Carmarthen, 7km west of
Conwyl Elfed. OS SN32NW GR32 35,25-28

Late June 1987 took me to a part of Great Britain I had never visited before, the gently rolling hills and valleys situated north of Camarthen. I was to spend a month there mapping the geological formations that defined the 'nose' of the central Welsh syncline.

The area I was to concentrate on centred around the village of Talog, which lay at the junction of the Afton Cywyn and one of its tributaries that flowed in from Pant-y-kkyn to the East. A once prosperous village boasting a water mill, a wheelwright and several cottage industries, it was now reduced to a single Post Office by the old stone bridge. The tracks to the Talog water mill could still be found when I last visited and on the opposite west bank of the Afton the remains of a forgotten cart track could be followed up to the remains of a saw mill at Pencaerau (though now over planted by a manageable coniferous forest). It was while following this wooded avenue that I first encountered the wild marauder of Talog.

The day was warm and sunny but heat had failed to penetrate the cool gloom beneath the forest canopy. As lunchtime approached I decided to break cover and try and find a field where I could enjoy the June sunshine. Holding my bearings was difficult due to the density of the foliage and several times I had to stop, sit and contemplate my location or simply back track.

When the terrain around me started to rise quite steeply I decided to recheck my position, not remembering a hill I thought that I had gone too far off course. Suddenly I heard the noise of someone or something moving around in the undergrowth. I listened and to my horror deduced that what was making the noise was both quite big and very quick, though thankfully some distance away. Whatever it was never broke cover and I did not feel inclined to pursue it.

The same thing happened a few days later on the edge of the wooded area by Penrhiwlas Farm. This time I was able to conclude that the thing was less than a foot tall but still quite large (it could disturb an area of ground cover three or four feet square!). This pattern of sightings (?) continued over the next few weeks and I was beginning to think that I was being followed by a rogue SAS belly crawler.

It was well into my third week in the field when the mystery was solved. Whilst walking between Pencaerau Mawr and Blaen-y-coed I spotted a series of rabbit holes in the field to the north of the road. To farmers, rabbits are a pest, but to a geologist they can be a blessing for in the area where a rabbit digs its home it burrows into the thick covering of boulder clay that blankets the local terrain throwing out tiny fragments of the underlying rock, giving the geologist a rough idea of what is buried directly beneath the clay. While collecting tiny shards of discarded claystone from the warren I heard the tell tale sound of my elusive pursuer coming from a patch of bracken. As I glanced towards the location of the noise I was stunned to see the thing break cover. 'It' was none other than a pack of several small, terrier like dogs, each one no bigger than a foot in size and from what I can remember, of various doggy forms and mottled hues. The shy creatures quickly went to ground when they realised that they were being watched, but I caught sight of them again on the far side of the valley dashing unconcernedly through a field of cows. Although I heard them after that, I never saw them again.

©Ian Hawthorne 1995.

EDITORS NOTE: The interesting thing about Ian's account is that it appears to deal with a small pack of feral dogs. Whereas feral cats are well known denizens of the British countryside, and whereas feral polecat-ferrets are also well established, accounts of truly feral dogs are much rarer.

In the south-west, for example, we have vague, and only semi substantiated records of three packs of feral (or semi feral dogs), and one of these is so steeped in folklore that it is unlikely to refer to a genuine creature rather than to a zooform phenomenon like the archetypal celtic black dog.

The three records we have are:

1. A pack of lurcher type dogs left running semi wild in a valley on the northern edge of Exmoor by a family of Ilfracombe poachers. These animals, if, indeed they exist, bear the same relationship to truly wild dogs as do the sheep and ponies of Exmoor. The puppies are, or so I have been told, collected in the spring to be kept as poachers dogs and watch dogs. The only evidence for this population is anecdotal and the sources are not 100% reliable.

2. The black dogs of southern dartmoor rumoured to be descendants of those liberated by retainers of Squire Cabel of Buckfastleigh in Tudor Times. Although some of the reports of black dogs from the Buckfastleigh area in our files appear to be of living creatures this provenancing is highly doubtful. I would suspect that these are zooform phenomena.

3. The best attested pack of wild dogs has, according to our informants (who are fairly reliable) been living wild on Haldon Hills since about 1990. They are the descendants of 'Convoy Lurchers')my title for the surprisingly homogenous 'thin dogs on strings' seen outside every DHSS office in the land and usually kept by Travellers. There was a travellers 'park up' on Haldon Hills for several years, and I have been told that when they were 'moved on' by the Police they left some of their dogs behind.

We have two slightly better attested records of single dogs living wild in the area, the most impressive of which concerned an animal which lived on Northam Burrows for several years during the last war until falling foul of a land mine.

It does, however seem that in most cases dogs find it far harder to live independently of their human masters than do cats or even ferrets, and so when we do receive a record such as this, which appears to refer to dogs who have gone against what they learned in 'The First Friend' from Kipling's 'Just So Stories', it is particularly interesting.

* * * *

Mystery Cats in Scotland during 1995

by Tom Anderson.

(Scottish Correspondent for the Centre for Fortean Zoology).

Dateline-September.
From 'Our man in Scotland'
standing in something punguent in a field.

Since the first sighting on the third of January, there has been a recorded big cat sighting between Inverness and Tayside approximately every eleven days.

Having checked the provenance of most, I would class them as reliable, none having sensational overtones.

The salient facts to date:

- None were sighted on moorland per se, but predominantly on farming land, in three cases crossing fields containing cattle.

- Their range contains both coniferous and broad leaved woodland providing an abundance of cover. High level pine-woods being in the minority.

- Sightings vary from early morning to dusk, but do not seem to influence the witnesses' opinion of colouring.

- Size varies from domestic dimensions up to a total of five feet in length and three feet high.

- Length of pelt starts with being comparable to a fox terrier and peaks at being comparable to a border collie. The latter in only one case so far.

- "Black Puma" type accounts for ninety percent of sightings. Two of these sightings included cubs (in one case four of them).

- Large Revack/Kellas type, black, two sightings, one with single cub.

- Tawny large Kellas. One coastal sighting carrying rabbit.

- Unique is the shaggy animal seen this month by a retired farmer near Oyne, Aberdeenshire. Having previously seen the standard puma he stressed the length of hair, (four inches), and the gait (long, fast pacing strides, ears flat and a 'cruel expression').

- The very volume of sightings in the Grampian area reinforce the theory that we are dealing with a native animal rather than an introduction.

- Anyone releasing 'exotics' would go to the high country of the north rather than relatively tame farming land with a total population of a quarter of a million.

- Their prevalence in the area must be related to the numbers of sheep and Roe Deer available, a milder climate to that further north and a huge fox population, (no organised hunting here, which until recently took the blame for all sheep predation.

- Like the eagle of the highlands, the fox is a scavenger and is quite likely to be dependent on cat kills in some circumstances.

- In any case, since the introduction of the 'Dangerous Wild Animals Act' was only six years ago, the diversification of the cats encountered preclude any evolutionary involvement.

* * * * * *

Cryptomicromammology in Europe.

by Patrick Brunet-Lecompte.

Introduction.

The discovery of a new species of small vertebrate is not the concern of cryptozoology as one would usually consider it. When it, however, stems from the work of a zoological discovery and takes place in a region as well known as Europe, cryptozoology could gain from such a discovery.

There are two main reasons that explain the significance of micromammals for the cryptozoologist. The way that micromammal fauna of a given region is catalogued, and the large separation of species amongst rodents and insectivores. The aim of this article is to prove the cryptozoological significance of micromammals by means of carefully chosen examples amongst the rodents and insectivores of Europe, being the part of the world whose fauna is best known.

Examples of Geographical distribution of populations and species.

Knowledge of the micromammals of a given region is acquired using two main methods:

i. Analysis of Owl Pellets
ii. Trapping.

The first method makes it possible to quickly identify a large number of prey, but is limited by the range of diet of the predator and above all the altitude at which it lives.

In Europe, the principal predator for the purposes of collecting pellets is the Barn Owl *(Tyto alba)* which never exceeds an altitude of 800-1200 metres. Consequently trapping remains the main method of researching micromammals in the mountains, although it is a far less profitable method of identifying numbers of specimens than collecting owl pellets.

Furthermore, trapping makes it possible to carry out interesting biological analyses (cytogenetics, protein genetics, etc...), which can make the distinction between sibling species.

The Snow Vole *(Microtus nivalis)*, in the Vercors.

A massif in the western French Pre-Alps, the Vercors is cut off to the west of the Massif Central by the Rhone Valley to the east, and the massif of the Chartreuse by the valleys of Drac and Isere. Knowledge of the micromammalian fauna of the massif has only recently been acquired.

Microtus nivalis

The first significant study was carried out in situ during 1964 collected 500 micromammals by trapping. No *Microta nivalis* were found among them. *(Brosset et Heim de Balzac 1967).*

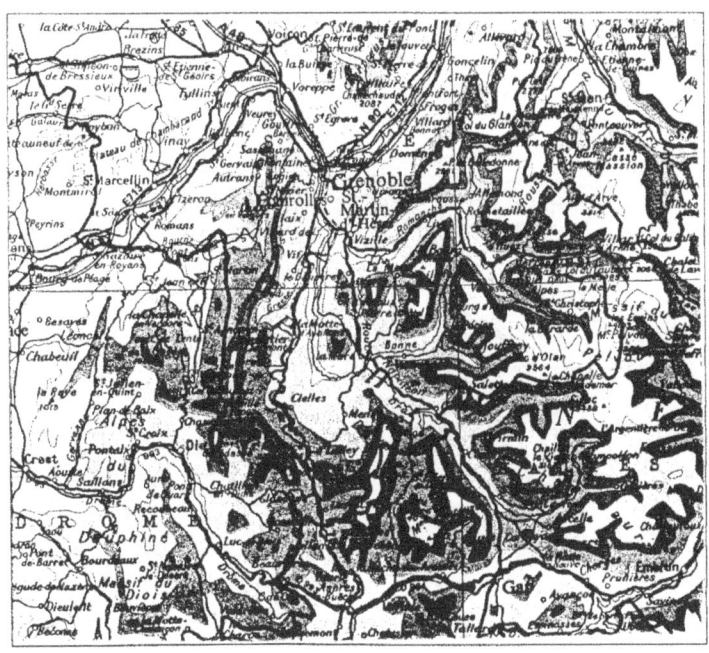

The Vercors

Although surprised by the presumed absence of the species otherwise present in the rocky biotopes of the inner Alps of the South of France, the authors appeared to accept its absence from the Vercors. Since then, additional research carried out in 1967 collected nine specimens in seven sites in the massif. *(Ariago et al. 1989)*.

The European Ground Vole, *(Microtus subterraneus)*, in Gironde.

Microtus subterraneus

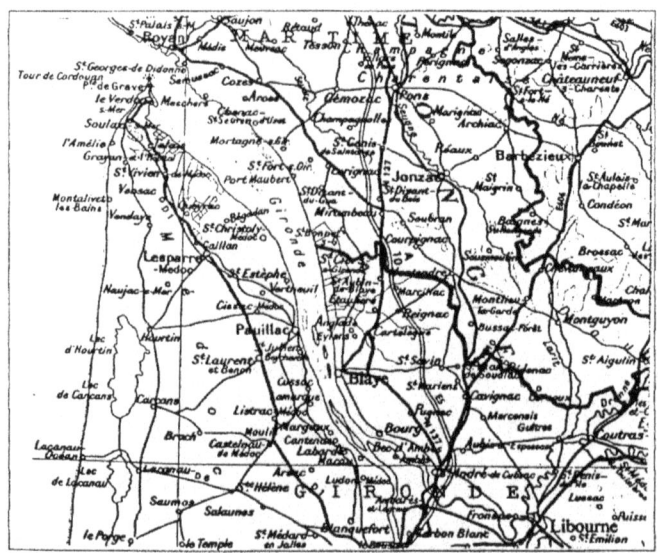

Gironde

The European Ground Vole, *(Microtus subterraneus)*, is a species that is not found in either the cold regions of boreal Europe or in the warm regions of mediterranean Europe. In France, the southern limit of its distribution gives through the Massif Central and the Charente (Fayard 1984).

The discovery of a population of this species in a batch of Barn Owl pellets from La Roquille in Gironde about 100 km south of the Charente *(Seronie-Vivien 1982)*, is worthy of some comment. The site of La Roquille did not seem to be a priori a favourable place for this species because of its southern position and low altitude (plain of Aquitaine). On the other hand, the analysis of five other batches of pellets in an area of 10-50 km around la Roquille showed no Microtus subterraneus in any of the 347 prey analysed (compared to 24 specimens in 137 prey in La Roquille) which brings to the fore the residual disposition of this population in Gironde. The discovery made by Seronie-Vivien (1982) which appeared in a local review is not even mentioned in the atlas of wild mammals of France *(Fayard 1984)*.

Examples of isolated species in Europe.

The comprehensive synthesis on rodents of Europe by Neithammer et Krapp (1982) mentions that several species have an isolated and restricted distribution. Among these are three species of the ground voles group.

- *Microtus felteni*, found only in southern Macedonia and northern Greece.

- *Microtus tatricus*, found in the tatrus mountains (Poland and the Czech Republic) and the Rodnei mountains in Romania (Flouzec 1985).

- *Microtus bavaricus,* found in Garmisch-Partenkirchen (Bayern, Germany) and in Biberwein (Tirol, Austria).

Chromosomal speciation.

Chromosomal speciation is a common means of speciation among micromammals. In Europe, one particular example is a good illustration of this speciation; the shrews of the group *Sorex araneus*.

The shrews of this group have a caryotype characterised by a sexual trivalent XY1Y2. Meylan (1964) found two different chromosomal types in Switzerland amongst populations which until then had been attributed to the one single species *Sorex araneus*.

Subsequent studies, *(Meylan et Hausser 1973, Hausser 1976)*, have confirmed the existence of two species that are morphologically very close but chromosomally different. *S. araneus* and *S. coronatus*. Since then cytogenic analyses in R-banding *(Volobouev et Catzeflis 1985)*, raise the problem of the existence of a third species in the group for the two populations of Valais (Switzerland).

The origin of the chromosomal transformations among these different populations could date

Sorex araneus

from after the last Quaternary glaciation, less than 10, 000 years ago. The speciation of the present species of the group *Sorex araneus* therefore seems to be a very rapid phenomenon lasting a maximum of a few thousand years.

The end of the last glacial period in Western Europe 10, 000years ago, and the geographical redistribution of the species which followed, lead us to make the assumption that chromosomal speciations exist in other groups of micromammals, particularly in mountainous regions. These speciations, favoured by the geoclimatic variations have already been proved in part among voles *(Niethammer et Krapp 1982)*, notably among the species of the subgenus *Microtus (Terricola)*, the European ground voles. *M (T.) subterraneus* is characterised by a caryotype 2n=52 to 54, whilst among *M. (T.) duodecimcostatus* a variable fundemental number is present according to the populations NF=72 to 76. The Alpine Ground Vole *M. (T.) nultiplex* is characterised by a very significant chrosomal polymorphism: 2n=46 to 48 *(Graf et Maylan 1980)*.

Conclusion.

The few examples presented here tend to show that there are still some existing species of rodents and insectivores yet to be discovered in Europe.

The work of the Cryptozoologist should include research in favourable regions, cytogenetic and electrophoretic analyses and analyses of the mitochondrial DNA of populations discovered in the same regions. This should contribute to such discoveries as the recent entry to the ranks of European fauna of the Alpine Wood Mouse *(Apodemus albicola)* by Heinrich (1992).

This has encouraged research into other isolated populations of wood mice in mountain areas such as the Pyrenees, the Apennines, the Iberian massifs and the Balkans.

The group of European ground voles which is already rich with discoveries during the 20th century could still have some surprises in store. More research is needed amongst isolated populations presently attributed just to the one species.

For example the populations of the sub group *subtrerraneus-multiplex* of the eastern edge of the Massif Central in France *(Fayard 1984)*, and the alpine form of the field vole *(Microtus arvalis)* which are subject to a morphological differentiation in relation to their fossorial behaviour. *(Spitz 1972, Le Louarn 1976)* together with a process of speciation which could be observed just as well among the Iberian populations of the Central System as among the populations of the Pyrenees of the *Microtus arvalis* group.

* * * * *

Three aspects of Durer's Rhino.

the Indian folk legend of the three blind men confronted by an elephant is well known and has been quoted at length in many fortean books, where it is used as an analogy to Fortean research. We have taken the analogy and are using it to approach, not an elephant but another giant, Asian land animal.

The engraving of the magnificent Indian Rhinoceros by Albrecht Durer is one of the best known icons of European historical Zoology. It has been reproduced on numerous occasions and has even been a LP sleeve, (the 'Mothers of Invention' spin off band 'Rhinoceros' in the late 1960's).

Here we present three views of this famous creature.

A trip to see the Rhino.

by Noella MacKenzie.

Marthe came back from the market one day with a new ballad for Ghislaine, and I about a strange animal from Africa called a RHI-NO-CER-OS. I told my school teacher about it and he said that he would look it up in his book of beasts, but I think that he forgot as usual. Maman said, the priest had told her that the animal itself was to be seen on an island by Marseille, and as we were going down to her sister's baby's christening, while Papa went with the flocks to the summer pastures, perhaps we would go and see it - but only if we were good, of course.

The day came to go, and we were off down the river. The summer heat was beginning and the roofs of the farms on the banks of the river seemed to tremble in the sun, while the cypress trees stood stock still. We came to the big city; I have never seen so many houses together, and so many people, all dashing about or seeming to be busy. Uncle is a grain merchant, and is always down at the harbour, arranging ships and cargoes. Auntie has plenty of fine clothes, and entertains a great deal, we always have lots of good food, all fancifully prepared when we come to visit our town cousins, and of course there is to be a great party to celebrate the naming of the new baby, who is to be called Denis, after our patron saint.

We went down to the harbour to get a boat to go across to see the Rhinoceros, but it wasn't so easy. There were crowds of people and the boatmen were all lined up by the quay, trying to get people to board their particular boat. There were booths selling food, pottery, wine, and all the sorts of things that you get at a fair, tumblers reciters and hundreds of people just wandering around. Whilst we stood there, I heard a sort of silvery sound in the air, and three people on horseback came through the crowd. There was a big woman, a burly man, and a little man dressed very shabbily; but they all rode magnificent horses. The man and the stout woman were wonderfully dressed, they must have come a long way, for they wore travellers' clothes, but all of the very best material, and their horses trappings were also rich and decorative. The sound had come from the lady's bridle, which was hung with silver bells. They dismounted and gave their horses into the care of an ostler nearby, but did not just throw the reins to him and ignore the beasts; they spoke to him and patted the horses and spoke to them also, watching all the time that they were properly handled.

Ghislaine was looking at them with eyes like saucers. Finally she said to Maman in wonderment, 'Are they kings and queens?' at which a woman on a food stall nearby laughed, and said, 'No, dear, they are just ordinary people like ourselves, but they live in the North, where they have fog and rain half the year, and snow and ice the other half, and they work hard to keep warm and make a lot of money; but they don't get much to eat except stringy mutton and sour cabbage, so every few years they come south to see the sun and enjoy some good food', and she looked with pride at her table, with its pies and pates and fruits all spread out on vine leaves on a white cloth.

Maman said that we ought to try and get a place in one of the boats if we were to see this remarkable animal, so she went timidly up to one of the men and asked him how much he would charge for two women and two children. He named a figure, and she turned away sadly. 'No children, it is too much to pay, we cannot take the voyage'. Ghislaine was very sad at this, and in fact so was I, but she, being a girl could let a few tears fall. At which the little man of the party of three came forward and said, in his clipped, foreign French; 'If this is not too much of a liberty, ma'am, may I pay for the children and their nurse?' Maman smiled at him - he appeared to be a gentleman despite his rather worn clothes - and said, 'That is most kind of you, sir, I will be pleased to accept, and with many thanks'. So we boarded the boat, he paid the boatman, and as we had what seemed to be the last spaces aboard, we soon set out. The vessel did not go very fast, as the man had got as many on it as he could, but the island was not very far from the quay and we were soon there.

But the same crowds that had been milling about in the harbour were also here, as the boats had been going to and fro all the time. There was a huge conglomeration round the animal, which was what we had come to see, of course, and a monk selling curios, which he assured the crowd were of real rhinoceros horn, and would keep away all sorts of ills. But when he saw the two richly dressed travellers, he immediately took a key from a bunch on a chain around his waist, and unlocked a box under his table. He took out something wrapped in velvet, and placed it on the table carefully. At first the two people didn't take any notice, or made out they didn't, and the monk just stood there, not attempting to praise the quality and virtues of this particular jewel, which was an ivory carving of two angels flying. Then the stout man went over to the table and glanced, apparently casually, at the carving, as though it was something quite pleasant to look at but not really valuable. Then the woman came over and first looked at the man, then at the carving, and you could see they were talking quietly between themselves; then they went up to the monk again and some money changed hands.The monk seemed pleased, wrapping up the carving carefully and handing it to the man who walked away with the woman without looking specially pleased about what he had bought. This had all happened in the case of a few minutes. Then the monk went back to crying his wares once more, Maman went up to the stall and bought a coral for our new cousin. Then we turned our attention to the serried ranks round the rhinoceros.

At the sight of all those backs, shoulder to shoulder, Maman got all timid and fluttery again. "I'm afraid we shall not see the animal after all, children", she said. But we hadn't reckoned with Marthe. She took one look at this wall of backs, squared her shoulders and said, "It was me that got you to see this Rhino", (she's a good cook but not much good at grammar), "And see the Rhino you shall! Just tuck in behind me with the children ma'am and sir", and using her elbows in the best market manner, she soon got us into the front row. The little man was not looking at the rhino, but at Marthe, with an expression of deep respect on his face.

Now we had really got a sight of this beast at last, everyone was quiet. He was simply enormous, with two horns sticking up on the top of his snout, which was broad, and a coat which hung in folds, like soft leather. Ghislaine just said "Oooh!" and that was all that we said, we just stood and tried to imprint the picture of the animal on our minds. I was mindful of the fact that the schoolmaster would expect me to write an essay about it when I was back home, and tried to think of words to portray something so new and strange.

Later, when we were going back in the boat, Ghislaine was sitting beside the old man and looking thoughtful. Then she turned to him and said (she really is cheeky), *"You didn't buy any of the expensive souvenirs, like the other two, did you?"* He looked at her seriously, and said gently, *"I have some children children at home to think about, and they have children of their own too"*. Ghislaine looked serious too, and said, *"They're better than all the jewel's aren't they?"* (she really is the limit). He replied, smiling at her kindly, *"Yes, really, I think they are"*.

We reached the quay and the little man went off to fetch his horse with the other two; he said that he would say goodbye properly in a moment. Then he came back and bowed to Maman and to Marthe, who was most pleased, and then turned to us. Ghislaine looked thoughtful for a moment then said, *"are you a real knight who comes to the aid of ladies in distress?"* He replied, *"Yes, I try to be"*. So she said, *"Will you be my champion against the wicked and the dragons?"* He smiled and said, *"Yes, Lady, I will"*. So she tried to look haughtily down her nose (snub), and handed him her kerchief (grubby) and said, *"Sir Knight, will you keep this in remembrance of me, strive to do right, champion the poor, and give succour to damsels in distress?"* He tried to bow low, but was very stiff; however, he swept his hat to the ground and said; *"Verily, lady, I will do as you ask, and defend your honour against all comers"*. He took the kerchief, tucked it into the band of his hat, bowed (as well he could), once more, and said; *"Farewell, most noble lady, and may God go with you"*. Ghislaine then blew him a kiss, and a lady nearby sighed and said, *"Charming"*, while the rest of the bystanders looked on in amusement.

The other two members of the party were already mounted, anxious to be away. So our knight said goodbye to us, mounted and departed, turning to wave. So they went away, back to the fogs and the boiled cabbage.

We went back to Uncle's house for the evening meal, and of course, everyone wanted to hear of our adventures. So we were late going to bed, and very tired - so much had happened in one day, that you needed a night's sleep to get used to it.

*****

In the 16th Century the first Rhinoceros to be seen in Europe was brought to an island in Marseille harbour (the Chateau d'if Island of Dumas's Monte Cristo) and shown to the public; thousands of people came to see it.

Editor's Note: One of the nicest thing about being an editor is that you can write your own game plan. As soon as I read the story above I wanted to publish it, and although it had little or no relevance to the work of the Centre for Fortean Zoology, I decided to include it in the yearbook. So first, I telephoned Clin....

Early Rhinoceroses in Europe.

by C.H.Keeling

The first Rhinoceroses to be exhibited in Europe were all of the magnificent Indian species (*Rhinoceros unicornis*) - one of the most impressive of all mammals.

One was shown in Pompeii in 61 B.C., another was exhibited at the games in Rome, organised by the Emperor Pompey a few years later, and another was displayed in the same city in 29 B.C. Incredibly, it is believed they were brought overland from India.

The first specimen of more modern times arrived at Lisbon on the 20th May 1515 as a present to King Emmanuel from the King of Gambay, via Alburqueque, the Governor of Portuguese India. By January of the following year, His Majesty had either grown tired of his pet, or he was seeking some favour of the Pope, as it was sent as a present to the latter, but unfortunately the ship conveying it was sunk in a storm in the Gulf of Genoa and the animal was drowned. This was the specimen from which the artist Albrecht Durer made his famous drawing.

The first individual to be seen in England was a young specimen that arrived here in October 1684 - when there were still living Dodos in the world, and wolves were found wild in Scotland. It toured the country for a year, but its ultimate fate is unknown.

Captain D Moyt brought a two year old female to London in 1741, (when there were still a few wolves in Scotland!) Within a year it had been sold to a showman in Holland, and over the next few years was shown in Holland, France, Germany and Austria; it was last heard of in 1749 when it was examined by the great French naturalist Buffon.

The London Zoological Garden's first Indian Rhinoceros (or any other species for that matter), arrived on the 20th May 1834 and died on the 19th September 1849 as a result of being accidentallly injured by an Asiatic Elephant, while on the 25th May 1864, the famous 'Jim' arrived at the same place, where he lived until the 12th December 1904, having lived in the capital for forty years, four months and seventeen days - which I believe is still the world longevity record for the species in confinement.

* * * * *

Editor's Note: Whilst the synchronicity between the date of the arrival of the first European animal of modern times, (and presumably the creature referred to in Miss MacKenzie's story), and the arrival of the first animal owned by the London Zoo, is worth noting there are other implications of interest to the Fortean Zoologist.

Close up of the shoulder of the rhinoceros pictured in Durer's engraving

Writing in his book, 'Extraordinary Animals Worldwide', (1991), Dr Karl P.N.Shuker discussed reports of rhinoceri of several species, including the Great Indian, which exhibited extra horns. He notes that Durer's engraving (which was, incidentally copied by Conrad Gesner in his 'Historae Animalium, Liber 1' (1551), shows an animal with a peculiar 'horn' on its shoulder.

According to Shuker, Dr Bernard Grzimek has suggested that this is a representation of *another abnormally horned specimen. It has also been suggested that it was "merely an excrescence developed by the rhino in question during its long confinement in the ship bringing it from India"...*

He goes on to note that Dr K.C.A.Schulz (*African Wild Life*, 1961), said that rough sores of a horny nature had been observed upon african Black Rhinos as well. As the corpse, has been lying on the bed of the Mediterranean for five hundred years now and has probably long since decayed, it seems unlikely that we shall ever know the truth!

Do Dinosaurs still roam the earth?

by Roy Kerridge.

Scientists never tire of reminding us that despite the assumptions of popular films, cartoon strips and trashy books, dinosaurs and human beings have never actually met. Cave men and dinosaurs, they claim, were separated by millions and millions of years. Certainly, most cave paintings show familiar animals such as bulls, horses and deer. Why, however, is the myth of man meets dinosaur so persistent? Could there possibly be something in it?

Consider this account by the naturalist and rain forest explorer, Ivan T. Sanderson. It is taken from his book 'Animal Treasure' (Macmillan, 1937), a book that inspired the youthful Gerald Durrell. Ivan and a companion were wading through a river in a remote part of Cameroon, West Africa, hunting for bats with Butterfly nets.

"*Suddenly George let out a shout:*

'Look Out!' and I looked.

Then I let out a shout also and instantly bobbed down under the water, because, coming straight at me only a few feet above the water was a black thing the size of an eagle. I had only a glimpse of its face, yet that was quite sufficient, for its lower jaw hung open and bore a semi-circle of pointed white teeth set about their own width apart from each other.

When I emerged, it was gone. George was facing the other way blazing off his second barrel. I arrived dripping on my rock and we looked at one another.

'Will it come back?' we chorused, and before it became too dark to see, it came again, hurtling back down the river, its teeth chattering, the air 'shh-shhhssing' as it was cleft by the great, black, dracula-like wings We were both off our guard, my gun was unloaded, and the brute made straight for George. He ducked. The animal soared over him and was at once swallowed up in the night".

Back in camp the two men found a group of African hunters with captured animals for sale. When Ivan described the creature he had seen, the effect was astonishing...

"'*Olitiau!*' somebody almost screamed" and with that the hunters drop everything and run away pell-mell, "*straight across country towards the village*", never to return!

Although Sanderson, now dead, never confirmed it, many people believe that he and his friend encountered a pterodactyl or flying dinosaur. As the nineteenth century adventurer and writer Trader Horn often remarked, Africa surely holds many secrets of large, undiscovered wild animals.

Editor's Note: Here it should, I think, be stated that many Cryptozoologists including Sanderson himself, believe that what Sanderson saw was a large and probably unknown species of bat, possibly allied to the Hammerheaded Bat *(Hypsignathus monstrosus)*, which although interesting is not as exciting as prospect as a surviving pterodactyl. It has been said that Sanderson himself considered that what he had seen was a bat.

Writing in his book *'In Search of Prehistoric Survivors' (1995)* Dr Karl P.N.Shuker discussed the episode in greater length and pointedly did not rule out the pterodactyl theory, although a similar prehistoric flying reptile, one of the rhamphorhynchoids seems to be a more likely candidate than a pterodactyl per se.

West of Lake Tumba, in the Congo, impenetrable swamps stretch for two hundred miles or more. African fishermen tell of an immense rhinoceros that roams the interior of this swamp. It is said to have a large, sweeping tail like that of a crocodile, a horn on its nose, and is capable of ripping an elephant to death. They say that it charges furiously at elephant intruders, who flee in panic. This does not sound very much like a rhinoceros. All known rhinos and hippos have short, stubby tails. It does, however, sound a great deal like a triceratops, or horned dinosaur.

When my three year old niece Omalara visited Port Lympne Wildlife Park, she greeted the sight of a baby African rhino with a glad cry of 'Sarah!'. She had mistaken the animal for Sarah, the baby triceratops in the full-length cartoon, Land Before Time'. Could not the same mistake have been made in reverse?

A typical Triceratops

Outside Africa, land dinosaurs may no longer exist, but there is some evidence that they survived in the Near-East until Old Testament times.

Nebuchadnezzar (604-562 BC), the ruler of Babylon, rebuilt the great Ishtar Gate, ceremonial entrance to the fortified city. Ishtar was the goddess of war and the patron of the gate, (which still exists today). It is a marvel of blue, glazed, brickwork, patterned with figures of great bulls and dragons.

The dragon was the symbol of Marduk, patron god of Babylon. Both bulls and dragons look very realistic. The dragons, however, do not resemble creatures of legend with wings and fiery breath. They look like straightforward dinosaurs, stalking along with tails, and lengthy necks held high. On each dragon nose there is a rhino like horn, very similar to those found on the fossilised bones of many species of dinosaur.

Do references to 'dragons' in The Bible refer to dinosaurs? Above all, what kind of creatures are Behemoth and Leviathan? They are described in the Book of Job by God Himself, as examples of his creative power. Biblical scholars glibly explain that Behemoth and Leviathan are respectively a hippopotamus and a crocodile. Turn to our Bibles, however, and we find that Behemoth too fails the hippo-rhino tail test.

"Behold now Behemoth, which I made with thee: He eateth grass like an Ox". (Job 40, v.15)

"He moveth his tail like a cedar: The sinews of his stones are wrapped together". (Job 40, v.17).

A Cedar is a mighty limbed tree, massive in girth and a tail that moves like it must surely be a dinosaur tail, not the tiny stub like appendage sported by the Hippopotamus. *:The sinews of his stones are wrapped together",* suggests reptilian scales, and not Hippo hide.

Leviathan has a chapter of Job to himself. (Job 41). He has a touch of dragon in him, as verse 21 describes a flame that *"goeth out of his mouth".* If dinosaurs are giant lizards, they may well flick long red tongues rapidly in and out and these could be mistaken for flames. Crocodiles, however, are carrion eaters with foul breath, and so could be the origin of 'smoke brathing monster' stories. At this moment a giant crocodile is terrorising fishermen in the Sea of Galilee, although it may prove, on capture to be a dinosaur.

"Cans't thou draw out Leviathan with a hook?" (Job 41, v.1).

"Cans't thou fill his skin with barbed irons? Or his head with spears?"

If he is a crocodile, yes, thou cans't. Africans using spears and barbed irons have killed thousands of crocodiles, and the Nile Crocodile is now and official 'endangered species'.

Leviathan, by contrast, seems to be a gigantic, invincible animal, a sea dweller whose very appearance strikes strong men with terror.

"Lay thine hand upon him, remember the battle, do no more. Behold, the hope of him is in vain: shall not one be cast down even at the sight of him?" (Job 41, v. 8-9).

"Who can open the doors of his face? His teeth are terrible round about". (Job 41, v.14).

"Round about" does not suggest a crocodile's long, pointed snout of a face. It sounds more like the fearsome jaws of a tyrannosaurus.

Many other verses refer to his scales, his teeth, and his un-crocodile like habit of raising himself out of the water. *"In his neck remaineth strength"*, verse 22 declares. A crocodile has virtually no neck, but dinosaurs are renowned for their neckiness.

"He maketh a path to shine after him. One would think the deep to be hoary". (Job 41, v.32).

EDITOR'S NOTE: It has also been suggested that these references are only semi descriptive and should be seen as semi-allegorical representations rather than subjective descriptions of a known or unknown animal. It has also been suggested that one or both creatures could be either a giant squid or a whale. The objections outlined in this article would therefore still apply.

Perhaps leviathan can be classed as a Sea-Serpent. If so he might be a long necked plesiosaurus.

Sightings of plesiosaur like sea monsters have been reported from all around the world, not only at sea but from freshwater lakes like Loch Ness, Lake Michigan and Lake Victoria. Most of these creatures differ from text book illustrations of plesiosaurs in that they have frills around their necks, and humped backs. Reconstructions of all prehistoric beasts are based on the evidence of fossil skeletons. If future non-human scientists were to reconstruct human beings on the same principle, their drawings would have no ears! Flesh decays and no-one can say with any certainty what prehistoric dinosaurs looked like. It has been suggested that plesiosaurs may have had inflatable 'camel-humps' on their backs as an aid to rising and submerging rapidly in deep water. If so, sea-serpents may be plesiosaurs. Camel humps are not visible on camel skeletons.

My last dinosaur story takes us to a very english spot. St. Leonard's Forest, near Horsham, Sussex. From the earliest of times this forest is said to have been haunted by dragons, monsters and serpents. St. George is said to have killed his dragon there, although Dragon Hill, in Wiltshire is the more well known site claimed for this legendary encounter, (which may possibly be based on a dim remembrance of human sacrifices to sacred crocodiles in India or Egypt).

In 1614, to everyone's surprise, a 'real' dragon seems suddenly to have appeared in St. Leonard's Forest. It was observed by many reputable citizens as it crashed around in the bracken. Most observers seem to have regarded it as a young, half, grown animal. Some tried to cast it as a traditional dragon with fiery breath, and supposed that bumps on its body might one day unfold into giant bat wings.

Eyewitness accounts were written down as part of an official enquiry into the dragon.

These show the animal, whatever it was, to have been nine feet long, thick in the middle and narrow at each end, which suggests a long tail and a long neck. It had black scales along the back, and red underparts. The neck which 'shootes forth' carried a white ring of scales. One sentence from the document, part of the Harleian Miscellany, sounds very much like an accurate piece of nature reporting:

"He is of countenance very proud, and at the sight or hearing of men or cattel, will raise his neck upright and seem to listen or to look about with great arrogancy".

Curiously enough the 'dragon' was said to leave a trail of slime everywhere it went. (See Job 41, v.32 again). In time, reports grew less frequent and the creature was heard of no more. In our own time, fossil footprints of a prehistoric iguanadon have been found in St. Leonard's Forest. These are millions of years old. A leaf eating dinosaur, the iguanadon is usually depicted as standing bolt upright when feeding, no doubt twisting its neck around 'with great arrogancy'. Could the seventeenth century villagers have seen a dinosaur ghost, one that left shining trails of ectoplasm? No, No. To admit as much would lead us away from science and into realms of pure fantasy!

* * * *

JANE BRADLEY 1961-95

Jane Bradley was not only a talented cartoonist but was also a great personal friend of mine. She contributed cartoons to the first five issues of 'Animals & Men'. She was killed in february 1995, and the post of Crypto-cartoonist has since been ably filled by the one andonly 'Mort', (by the way this pseudonym has nothing to do with Terry Pratchett, but is short for 'Mort D.Arthur' a pun of such nobly horrific proportions that it deserved to be enshrined for posterity).

When Jane died we had a number of cartoons and half finished sketches which we had not printed. These, with here other cartoons of a cryptozoological nature are gathered here together as a tribute to a very special lady. JD.

CRYPTO COMMENT: Noone has been too impressed with some of the more gruesome Crypto news stories that have been circulating recently. Cartoon by Jane Bradley

As Glastonbury Festival looms on the horizon-even the cryptids are getting in on the act....

Was an Onza shot in early 1995?

by Dr Rafael A. Lara Palmeros.

C.E.F.P Mexico.

(Mexican Representative for the Centre for Fortean Zoology).

The Onza is a Mexican cat, about the size and colouration of a puma (f.concolor), but it is said to be more aggressive, longer legged, more markedly gracile and rarer.

The first recorded mention of the Onza occurred in the 1700's. Father Ignaz Pfefferkorn was posted to Sonora in 1757, and worked with the Opata, Pima and Eudebe indians for eleven years, until the Spanish Crown expelled all Jesuit Missionaries from Mexico. He spoke of:

"...the animals which the Spaniards call Onza is in shape almost like the animal (puma) just described. However, it has a longer body, which is also noticeably thinner and narrower (.....) It is not so timid as the (puma), and he who ventures to attack it must be well on his guard". [1].

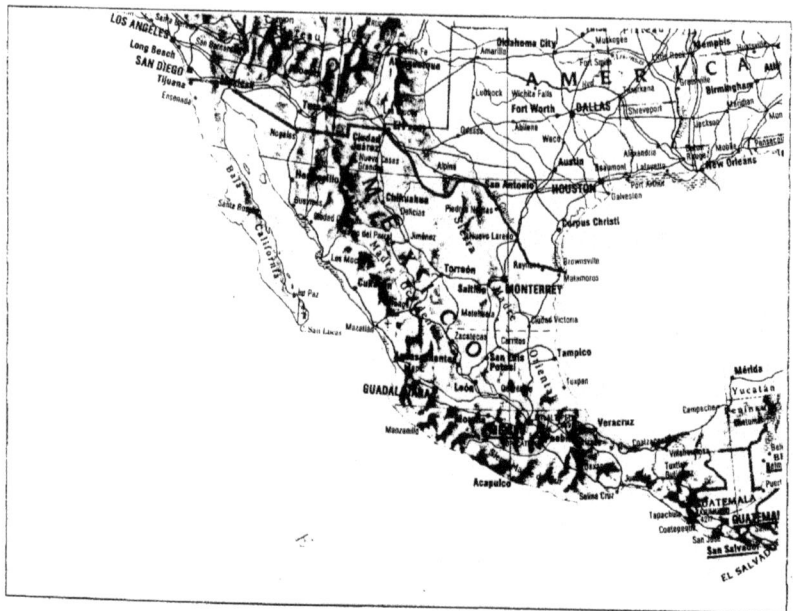

Mexico

On April 15th 1995, an Onza was shot by Raul Jiminez Dominguez, a Mexican rancher in a valley behind the Cerro del Perico (Parrot Mountain) in Sinaloa State [2]. Upon inspection, the hunter concluded that the cat was not a jaguar, and it looked different from a puma.

Mr Raul Jiminez spoke to his father who recommended getting the specimen to Mazatlan where it could be frozen. The time between the death of the animal and the freezing of the body was calculated as being about fourteen hours, and the specimen was in excellent condition when it was frozen.

The cat, a male, had a remarkably gracile body, with long, slender legs and a long tail. The ears also seemed very long for a puma (about 100 mm), and small horizontal stripes were found on the inside of its fore-limbs, which, as far as has been determined to date, are not found in the puma. Its total length was 184 cm. The tail was 72 cm. It weighed about 26 kgs.

The same day two Biologists of Universidad Nacional Autonoma de Mexico, Bi61. Arturo Duran Fdez y Biol.Antonio de Jesus Dannaure R. took tissue samples from the kidney, liver and muscles. The samples were preserved in small vials for electrophoresis analysis at the University of Mexico [3].

Since then we have received no reports about the results of the analysis. What happened? I don't know!

REFERENCES.

1. 'The History of Sonora - a cultural perspective' by Dr Raymondo Velazquez. pp 33-34. Editorial del Gobierno de Sonora. 1992.

2. 'El Clarin'. Newspaper. Morning edition. April 16th 1995. Mazatlan. Sinaloa State.
'Excelsior' Newspaper. April 16th 1995. Mexico D.F.

3. 'El Clarin'. Newspaper. evening edition. April 16th 1995. Mazatlan. Sinaloa State.

* * * * *

The story of a strange Fish
Quasi-fortean aspects of the discovery of the Coelecanth
by
Jonathan Downes

The Crossopterigian fishes, or coelecanths had flourished over two hundred million years ago but they had seemingly dies out overthe succeeding millenia, and the most recent specimens known were still sixty million years old.

Just before lunch on the 22nd December 1938, Marjorie Courtenay-Latimer, the curator of a small museum at East London about half way between Durban and Cape Town on the Indian Ocean coast of South Africa, received a telephone call from the manager of a local commercial fishing company to tell her that one of their trawlers had just docked, bringing in a pile of fishes that she might care to examine. She had spent a long time in attempting to educate her fellow townsfolk in natural history, and she had always encouraged people to keep her informed of situations which might lead to her obtaining some new specimens for her museum.

She was quite pleased with the telephone call from Irwin and Johnson, the owners of the fishing boat because they had previously been quite useful to her in allowing their captains and crew to collect potential Ichthyological specimens for her. When she arrived at the wharf, she found a pile of dead sharks waiting for her, but then out of the corner of her eye she saw something very different, hidden beneath them.

The crew disentangled this particular dead fish from the pile of sharks, and Miss Courtenay-Latimer was astounded to realise that what she was looking at was unlike anything she hadever seen before. It was five feet long and weighed 127 pounds. It was mauvy-blue in colour and was faintly flicked with white spots, but what was particularly startling was the set of remarkable fins. It had a pair each of dorsal, pectoral and pelvic and a single anal fin. Like the body fins of most modern bony fish, the first dorsal fin was borne directly upon the fishes body and contained a number rays ariusing from the very base of the fin giving it a fan like structure. The rest of the body fins, were however borne on a muscular lobe from which the rays arose giving the fins the appearance of short, primitive stumpy legs.

The tail fin was even stranger. Whereas in most species of modern fish the backbone ends at the base of the tail fin which in turn comprises two distinct lobes-the upper one directed upwards and backwards, the lower one directed downwards and backwards. In contrast the tail of Miss Courtenay-Latimer's fish was pointed with the backbone of the fish running right through to thepoint. It was superficially similar to the tail of modern lungfishes and eels, but in a startling departure from the anatomy of any known species of modern day fish, the East-London specimen had a third lobe sandwiched between the upper and lower lobes yeilding a tripartite tail. The other anatomical anomaly of note was the extreme oiliness of the fish which was still exuding a foul smelling oily exudate as she bundled the dead fish into the back of a taxi, whose dfriver was very unwilling to take his strange antedeluvian passenge back to the East-London Museum.

Miss Courtenay-Latimer learned that the trawler, under the command of Captain Goosen, had been trawling about eighteen miles south west of East-London, and they were about three miles off shore near the mouth of the Ohalumna River, heading slowly to the north east when the mysterious fish had been caught at a depth of about forty fathoms as part of a catch weighing over three tons.

Miss Courtenay-Latimer arranged for a local taxidermist to preserve the specimen and she gave strict instructions that the soft parts of the fishes interior were to be preserved separately. Unable to identify the fish herself, she wrote a long letter containing a description of this new and exciting specimen to South Africa's leading Ichthyologist, Professor J.L.B.Smith at Rhodes University College in Capetown. The events that followed were almost farcical in the number of 'stupid little' things that went wrong, (which is not at all uncommon within Fortean Zoology), and with the benefit of historical hindsight it is suprising, not only that the fish ever got identified at all but that the whole episode has become one of the most revered sections of the iconography of 20th Century Zoology.

Indeed the whole history of both cryptozoology and the wider field of fortean zoology is littered with such episodes of such farcical silliness, that it has prompted some of the more flamboyant of crypto-investigators, such as the legendary fortean Tony 'Doc' Shiels to invent their own bizarre methodology, ('Doc' is a confirmed believer in what he calls 'Surrealchemy').

The first problem occured when Miss Courtenay-Latimer's letter arrived at the professor's house. He was away on holiday, four hundred miles away at Knysia, a small town near the coast about halfway between Cape Town and Port Elizibeth. Due tothe congestion of the post over the Christmas holiday period the letter didn't reach him for ten days.

The second problem lay with the taxidermist himself. The story differs according to which account you read. Either the taxidermist failed to preserve the internal organs properly, or, according to an earlier account, Miss Courtenay-Latimer did not have the fish preserved straight away, hoping instead that Professor Smith would get her letter and come to examine the corpse relatively quickly. Either way most of the fleshy parts of the body deteriorated beyond recognition, and by the time Professor Smith telephoned Miss Courtenay-Latimer in early January all that remained of the fish was the taxidermically preserved cast and skin, the skull, and some portions of the spinal column and pectoral girdle.

More unforseen events intervened and it was not until the 16th of February that Professor Smith was able to examine the stuffed fish in person when he discovered that his initial hunch upon seeing Miss Courtenay-Latimer's sketches had been correct and that the fish was indeed a surviving Crossopterigian, and he named the creature Latimeria, as an affectionate tribute to the South-African spinster who had been so instrumental in its discovery.

Professor Smith believed that the type specimen of Latimeria was probably a stray and that the most likely habitat for this prehistoric survivor would be moderately deep water in the Mocambique channel between Madagascar and the African mainland. His plans for an expedition to the area fell through but undaunted he prepared a leaflet in French, English and Portugese giving a brief account and photograph of the fish and offering a reward of a hundred pounds for each of the first two specimens caught. Thousands of these leaflets were distributed to centres of population in the Western Indian Ocean, and Smith and his wife made many journeys up and down the East African coast to impress personally on local authorities the importance of the search but even it took nearly fourteen years for the next step in the drama to unfold.

On December 20th 1952, two days before the fourteenth anniversary of the discovery of the type specimen, Ahmed Hussein, was fishing of the island of Anjouan in the Comoro Archipelago, a French Colony off the north eastern coast of Madagascar and between Madagascar and Mocambique, when suddenly his line and fish baited hook were jerked out of his hand. After a titanic struggle he managed to land a hundred pound fish of a kind that he had never previously seen, and thus the second coelocanth was secured for science. Unfortunately, once again the gremlins of crypto-investigative methodology once again took a hand and as the only way that Ahmed Hussein managed to subdue the creature was by battering its head to a pulp with a piece of wood, the brain of this priceless specimen was practically destroyed.

Hussein, of course had no idea of the value of his catch but he merely saw it as 100lb of portable protein, and he took it to the local fish market for sale, and it was just about to be cut up for food when a passer by recognised the fish as being the subject of Professor Smith's leaflet. The carcase was then carried across the island to Captain E.E.Hunt, the skipper of a trading schooner who ordered the carcase salted for preservation.

Unfortunately the deck hands to which Captain Hunt had entrusted the task of salting the dead fish were over enthusiastic and got a little carried away with their task of cutting the fish open for salting and thus more of the internal organs were severely lacerated which meant that when Professor Smith finally got hold of the specimen more vital studies were unable to be carried out. Captain Hunt then obtained some formalin from a local doctor and liberally injected it into the dead fish.

The catalogue of misadventure was not yet over. Once again Professor Smith suffered unforseen obstacles in getting to the Comoros to retrieve the second specimen. There seemed every chance that even after his fourteen years of effort and despite Hunt's enthusiastic if over zealous work on attempting to preserve the carcase Smith would be unable to get to the specimen in time to stop the tropical heat turning his precious coelecanth into a nasty mass of putrefied flesh and fish oil. Then, as sometimes happens, fate in the shape of Dr Daniel Malan, the President of South Africa stepped in and events started to look up as Smith flew to the French colony courtesy of an especially diverted aircraft from the South African Air Force. When he got there he was happy to find out that despite the heat and the drastic lacerations to the fishes body that it was suprisingly well preserved and that finally he had the first internal organs of a coelecanth for study.

Since 1954 there have been a number of new specimens and at least one film has been taken of a living specimen in its natural habitat 600 feet beneath the sea.

The discovery that coelecanths did not die out millions of years ago as had been previously thought is important for a number of reasons. DNA analysis has confirmed that the crossopterigian fishes are the closest living relatives to the early amphibians and the the discovery has thus shed much light on the initial evolution of land animals. Professor Smith's initial theory that the coelecanth was endemic to the area surrounding the Comoro Islands and that the first fish from East London was a straggler has been borne out although the 1954 specimen was also atypical. It had so many morphological differences with the number and positioning of its fins that Smith initialy described it as a seperate species, calling it Malania in honour of the South African politician whose last minute loan of a military aircraft allowed him to secure the first relatively complete fish for science.

The most important implication however, and the reason why I have written a lengthy description of the discovery of a 'known' species when I spend most of my time on the trail of animals whose nature is far less certain, is as an object lesson. If one primitive creature can survive for sixty million years unnoticed then why not others? It is also an object lesson in how the forces of fate tend to act to confound your every move whenever one is actively searching for unknown creatures. 'Doc' Shiels calls it 'psychic backlash', another well known pundit within the field described it as 'the gods of crypto-investigative theory buggering us about again' and yet another puts it down to 'bad karma', and I personally think that it is because moist significant discoveries in our field are made by amateur's who have neither the resources nor the training to avoid pitfalls that the 'professionals' would not usually fall into, but nevertheless have the insight, and, dare I say it, the 0courage to follow the more untrodden ways in search of new adventures. Whatever the cause these continual stupid accidents do happen, and they have happenned to all of us who stray from the preset path in search of the more unusual aspects of mother nature.

* * * * * *

INDEX TO ISSUES 1-7 OF *'ANIMALS & MEN'*

(Note: The issue Number appears first in brackets)

INDEX OF AUTHORS AND MAJOR EYEWITNESSES CITED

ANDERSON T	(6) 29; (6) 31
ARNOLD N	(6) 13-15
BLASHFORD-SNELL, J	(3) 15-19
BLYTON E	(1) 23
BORD J	(7) 36
CARTER R.A	(4) 22-3
CHAPMAN B	(2) 9-12
DAVIS M	(1) 24
DE-SARRE F	(6) 23-25
DOWNES A.S	(1) 25; (1) 27; (2) 21; (2) 24-25; (3) 30-31
DOWNES A.S & DOWNES J	(1) 9
DOWNES J	(1) 10-18; (1) 24-5; (4) 17-20; (5) 22; (5) 35-36; (5) 39; (6) 11-12; (6) 15-16; (7) 13-16
FRASER M	(7) 28-30
"GAVIN"	(6) 16-18
GRAYSON M	(5) 32
HUTCHINGS R	(3) 27
KEELING C.H	(1) 24; (6) 32-33; (7) 37-38
KERRIDGE R	(5) 13-17
KINGSHOTT J	(5) 17-21; (6) 4
KRANTZ G.S	(7) 16-18
LEADBETTER S	(2) 13-6; (5) 28-29
MALORET N	(1) 19-20
MANNETJE M	(3) 29
MORGAN N	(3) 27; (5) 33
NAISH D	(7) 19-27; (7) 37
NATHAN P	(4) 21-22
NICHOLSON M	(4) 24

NIXON N	(6) 22; (7) 41
'PATERFAMILIAS'	(2) 19;
PARSONS S & DOWNES J	(3) 19
PETROVIC W	(1) 23
PLAYFAIR M	(7) 38
PRINGLE A	(4) 9-11; (6) 30
RETIRED COLONIAL OFFICER	(3) 26
SHIELS T	(6) 19-20
SHIPP S	(3) 12-14; (5) 26-27
SHUKER Dr. K.P.N	(4) 12-16; (5) 22-25
SHUKER Dr. K.P.N & SHIPP S	(6) 26-29; (7) 30-35
SORENSEN E	(5) 31; (6) 21-22; (6) 30
STEBBINGS S	(2) 28; (5) 34
STEPHENS A	(5) 34
STOCKER G.M.	(4) 22
THOMAS L	(4) 21
THURGAR H	(1) 24
WILLIAMS J	(1) 6-7; (1) 8-9; (1) 24; (1) 26; 920 17-18; (3) 7-8; (3) 20-25; (3) 29; (4) 25

INDEX OF SUBJECTS

(In the interest of brevity latin names have only been used where there is no common name, or where it seemed otherwise appropriate).

Abu Sotan	(1) 26
Agogwe	(1) 26; (6) 23-5
Ahool	(1) 26
Albino Birds	(2) 24; (3) 31
Albino Buffalo (see Buffalo)	
Albino Cormorant	(3) 31
Albino Puffin	(3) 31
Alien Big Cats	(1) 4-6; (2) 4-5; (3) 7-11; (3) 20-25; (4) 5-6; (4) 28; (5) 4-6 (5) 13-17; (5) 17-21; (5) 31; (6) 4; (6) 5-6; (6) 13-14; (7) 7-8; (7)13-16
Alligators	(2) 6; (7) 9
Amarok	(1) 26
Ambulocetus natans	(1) 7
Ameranthropoides loysii	(1) 26
Anaconda	(1) 26
Andean Wolf	(1) 26
Angeoa	(1) 26
Ape Sex	(6) 10
Arassas	(1) 26

Archaeopteryx	(7) 36
Artrellia	(3) 19
Atlas Bear	(1) 26; (2) 27
Australopithecus	(6) 23-25
Aypa	(1) 26
BHM (see Bigfoot; Orang Pendek; Hominoids of Africa; Yeren; Wildman etc)	
Baboons	(5) 10; (6) 10
Badigui	(2) 26
Bagenza	(2) 26
Bai-Xiong	(2) 26
Bakanga	(2) 26
Ban-Manush	(2) 26
Bantams	(2) 19
Baraboedaer Beasts	(4) 28
Barmouth Monster	(2) 26
Batsquatch	(5) 39
Batutut	(2) 26
Bear	(1) 6; (1) 26; (2) 18; (2) 26; (2) 27; (4) 7; (7) 4; (7) 6
Bear in Dolomites	(7) 6
Bear in Oxfordshire	(2) 18
Beast of Bala (lemur)	(5) 10; (6) 30
Beast of the Alps	(7) 6
Beast of Chiswick	(3) 7
Beast of le gevaudan	(2) 26
Beech Marten	(1) 10-18
Bigfoot	(2) 7; (2) 26; (3) 35; (4) 31; (6) 36
Bigfoot (Malaya)	(5) 8
Bird Euthanasia	(1) 25
Birdmen	(4) 28; (4) 31; (5) 39
Birdmen (Also See Owlman)	
Black Dog	(3) 12-14
Black Swan (aberrant)	(3) 30
Black Throated Thrush	(1) 27
Blashford-Snell, John	(2) 8; (3) 15-17; (6) 11-12
Blythe's Pipit	(4) 27
Boar	(1) 6; (5) 13-17; (6) 7
Bondegazou	(3) 4
Booaa	(3) 29
Boobrie	(3) 29
Bradley, Jane Obituary	(5) 36
Bray Rd. Beast	(4) 28
Brentford Griffin	(2) 26
Bruckee	(3) 29
Buffalo, sacred	(3) 7
Bunting, Yellow Browed	(3) 31
Bunyip	(3) 29
Burrowing Owl	(3) 31

Buru	(3) 29
Butterflies	(5) 7
Cabyll-Uisge	(6) 34
Cadborosaurus (Caddy)	(6) 34
Caiman	(7) 9
Cait Sidh	(6) 34
Camahueto	(6) 34
Camoodi	(6) 34
Capercaillie	(6) 39
Carp	(6) 31
Cat (Alien Big) see Alien Big Cats	
Cat Bornean Reed	(3) 5
Cat Forest Wild	(5) 31
Cat Golden	(2) 7
Cat Hybridisation	(5) 31
Cat Jungle	(5) 31
Cat 'King' Kellas	(6) 29
Cat Sabre Tooth	(6) 21-2
Cat Sand	(6) 9
Cat Scottish Wild	(3) 5
Catfish	(1) 6-7
Cayman	(2) 6
Chalicothere	(6) 32-33; (7) 37
Challicum Bunyip	(6) 34
'Champ'	(6) 34
Chequered Skipper	(5) 8
'Chessie'	(6) 34
Chimpanzee	(7) 8-9
Clear lake creature	(1) 6-7
Cormorant (albino)	(3) 31
Crabs	(1) 6
Creature Falls	(1) 19-23
Crocodile	(7) 9
Crocodile, centenarian	(6) 7
Crocodiles in New Britain (see Migo)	
Curlews, Eskimo	(2) 25
DNA Reconstitution	(5) 8
Dakataua (see Migo)	
David's Tree Partridge	(1) 27
Dogs (Black)	(3) 12-14
Dogfish (in Mersey)	(1) 6
Dogfish (2 headed)	(2) 29
Dolphin, Gangetic	(6) 12
Dyke, Nick and Sally	(2) 4
Duck, Merganser	(3) 30
Duck, Muscovy	(2) 25
Duck, 'Super'	(6) 39

Dung	(5) 12
DURRELL, Gerald	(5) 35
Eagle, Bald	(6) 39
Eagle, Philipines Monkey	(7) 11
Eagle, Sea spp	(6) 39
Elephant (cartoon)	(5) 12
Elephants (Humped)	(2) 8; (6) 11; (7) 37
Emu (4 legged)	(1) 27
Essex (washed up bodies)	(4) 25
Feral Children	(5) 9
Fighting Cocks	(5) 30
Fish Falls	(1) 22; (1) 23
Flamingo	(6) 39
Fox, Arctic	(3) 7
Fox	(2) 5; (2) 31
Fox attacks humans	(2) 5; (2) 31
Frog American Bull	(7) 9
Frog Falls	(1) 19-23
Frogs	(1) 19-23
Frogs (coloured) see Golden Frogs	
Frogs (3 legged)	(3) 11
Geese	(2) 24; (5) 30
'Giant Rabbits'	(1) 24
Gilbert's Potoroo	(5) 7
Goat, Hermaphrodite	(5) 12
Golden Cats	(2) 7
Golden Frogs	(1) 19-23; (4) 24; (5) 34; (6) 8
Grass Snakes (Giant)	(1) 24
Grass Snake (2 headed)	(3) 6; (4) 8;
Guinea Fowl	(3) 30
Gulls attack man	(3) 31
Gull, Ross'	(1) 27
Hairy Hands of Dartmoor	(5) 26-27
Hamsters	(5) 12
Hares	(6) 9
Hawk Moth (Sphurge)	(3) 7
Hedgehog ('Flying')	(2) 7
Heuvelmans Bernard	(2) 7
Hoaxes	(2) 7
Holy Goat	(4) 4
Hominoids of Africa	(6) 23-5
Homo erectus	(6) 23-5
Homo habilis	(6) 23-5
Hoppy the Toucan	(1) 27
Horse Przewalski's	(6) 8
Humped Elephants	(2) 8
Ibis, Japanese Crested	(1) 27

Japanese Crested Ibis	(1) 27; (4) 27
'Jelly' from sky	(4) 6
Komodo Dragon	(7) 11
Lake Dakataua (see Migo)	
Lake Monster (Argentina)	(3) 11
Lake 'monster'(Cheshire)	(6) 6
Lake Monster (China)	(5) 11
Lake Monster (Wales)	(6) 6
Large Copper (Butterfly)	(5) 7
Lemur Pygmy Mouse	(7) 5
Lemur Ruffed (See Beast of Bala)	
Leopard, Persian	(6) 10
Leopard references in ABC reports see 'Mystery Cats' etc	
Little Owl	(3) 31
Lion	(2) 7; (5) 12; (7) 8
Lion, Asiatic	(6) 10
Lizards (Common)	(2) 22-23
Lizards (Green)	(2) 22-23
Lizards (Iguana)	(6) 8; (7) 9
Lizards (Monitor)	(3) 6; (3) 15-17; (6) 6; (7) 11
Lizards (Mystery)	(2) 28
Lizards (Sand)	(2) 22-23
Lizards (Viviparous)	(2) 22-23
Loch Lomond 'monster'	(7) 38
Loch Ness Monster	(1) 4; (2) 13-16; (2) 23; (3) 27-28; (4) 22-23; (5) 28-29; (5) 32-33; (6) 7; (6) 31; 7; (7) 6; (7) 16-18
Loy's Ape	(1) 26
Luth	(2) 29
Lynx, Pardel	(2) 7
Macaque Rhesus	(5) 10
Man Beasts (New Zealand)	(3) 11
Manatees	(4) 12-16
Megamouth Shark	(5) 11
Meinerzhagen Richard	(2) 24
Mermaids	(1) 7
Migo	(2) 8; (3) 28; (4) 17-20;
Monitor Lizards	(3) 6; (3) 15-17; (6) 6
Moose	(6) 7
Muntjac, Black	(6) 10
Muntjac, Giant	(2) 6;
Muntjacs in UK	(4) 4
Mystery Cats	(1) 4-6; (2) 4-5; (3) 7-11; (3) 20-25; (4) 5-6; (4) 28; (5) 4- (6) 4; (6) 5-6; (6) 13-14; (7) 7-8; (7) 13-16
Nandi Bear	(6) 32-33; (7) 37

Natterjack Toad	(6) 8
Norfolk Snarlegow	(3) 7-8;
Night Parrot	(2) 21
Octopi	(5) 7
Okapi	(4) 21
Orang Pendek	(5) 7
Owls	(3) 31
Owlman of Mawnan	(6) 15-16; (6) 16-18; (6) 19-20; (7) 36
Panda, red	(7) 11
Partridges	(1) 27; (2) 25
Pardel Lynx	(2) 7
Parrot, Amazon spp	(5) 30
Parrots (& whisky)	(3) 30
Parrot, (Mr Porat's)	(2) 25
Parrot, night	(2) 21
Parrot, (swearing)	(4) 27
Peccaries	(1) 6
Penguin	(1) 9
Pigeons (on dope)	(2) 25
Pine Marten	(1) 10-18
Piranha	(5) 11; (7) 10
Polecat	(1) 10-18; (2) 28
Potoroo	(5) 7
Puma (not ABC)	(3) 11
Puma (Eastern)	(4) 7; (7) 16-18
Pygmies in Australia?	(5) 10
Quail	(1) 25
Rabbit ('flying')	(2) 7
Rabbit ('giant')	(1) 23
Rabbit ('nests')	(3) 27
Ransome, Arthur	(4) 23-22
Rats	(4) 8; (6) 9
Red Flanked Bluetail	(3) 31
Rhesus Macaque	(5) 10
Rhinoceros Javan	(1) 28-9
Rhinoceros Sumatran	(1) 28-9
Ross' Gull	(1) 27
Rough Legged Buzzard	(4) 27
Ruffed Lemur (in Wales)	(5) 10; (6) 30
Sable	(1) 10-18
Sasquatch	(7) 16-18
Scorpions	(3) 26
Seabirds	(1) 27
Shark	(1) 6; (5) 11; (6) 30; (6) 31
Shark, Great White	(5) 11
Shark, Megamouth	(5) 11
Shar, Sail-finned rough	(7) 10

Shrimps, Coal eating	(2) 6
Sloth, Aquatic	(6) 9
Sloth, Giant	(1) 8-9
Snails	(4) 7
Snakes	(1) 24; (3) 6
Snake, 'horse headed'	(7) 4
Sphurge Hawk Moth	(3) 7
Spiders	(2) 7; (4) 7; (5) 7;(7) 10
Spiders (fighting)	(2) 7
Sponge	(6) 10
Stegodon	(2) 8; (6) 11-12
Stone Marten	(1) 10-18
'Super Daddy Longlegs'	(4) 4
'Swallows and Amazons'	(4) 21-22
Tatzelwurm	(2) 20-21
Thrush, Black Throated	(1) 27
Thylacine	(2) 4; (4) 9-11
Tiger, Bengal and unspec(1fied subspecies	6; (5) 12; (5) 15; (6) 10; (7) 11
Tiger, Siberian	(6) 10
Tiger, Sumatran	(6) 10
Tizzie Wizzie	(2) 7
Tortoise, Mary River	(7) 4-5
Toucan	(1) 27
Tree Kangaroo	(2) 4
Tuatara	(7) 5-6
Turtle (Leathery)	(2) 29
Turtle, marine spp	(7) 11
Turtle Red Eared	(7) 10
Vu Quang	(1) 30; (2) 6; (3) 4; (3) 32; (4) 4; (4) 7; (6) 10
Vu Quang Ox	(3) 32; (4) 7
Vultures	(4) 6; (5) 30
Wallabies	(1) 24; (4) 4; (7) 9
Whales (stranded)	(4) 8
Whales (carcass mistaken for sea monster)	(4) 8
Whales, mystery	(7) 19-27
Wild Boar	(1) 6; (5) 13-17; (6) 7
Wild Man	(2) 9-12
Wild Man (Chinese)	(4) 4; (5) 9; (7) 5
Wolf (Andean)	(1) 26
Wolf-like apparition	(7) 28-30
Wolverine	(7) 9
Woodpecker	(6) 39
Woodwose	(2) 9-12
Wooly Jumpers for seabirds	(5) 30
Worm (Giant)	(3) 19

Xenoperdix unzungwensis (1) 27
Yellow Browed Bunting (3) 31
Yellow Browed Warbler (4) 27
Yeren (see Wild Man Chinese)
Zooform phenomena that resemble cryptids (7) 28-30

INDEX OF ARTICLES

(Note: Individual entries in 'Newsfile' or 'Nervous Twitch' (by Alison Downes) are only indexed below if they are substantial in length. in indexing words like THE or A have been ignored. Queries in the 'HELP' section have been mostly ommitted as have book, magazine and video reviews. Individual letters have been included and are cross referenced by author. 'Now Thats what I call Crypto' which ran from issue 6 is written by Neil Nixon)

A-Z of Cryptozoology part 1 by Jan Williams (1) 26
A-Z of Cryptozoology part 2 by Jan Williams (2) 26
A-Z of Cryptozoology part 3 by Jan Williams (3) 29
A-Z of Cryptozoology part 4 by Jan Williams (4) 28
A-Z of Cryptozoology part 5 by Jan Williams (6) 34
Ain't nessie-cerally so by Petrovic (2) 16
Alpine Enigma by Roger Hutchings (2) 20-1
Another fine NESS .... (letters) from Nick Morgan, and Martien Mannatje
The Bear Facts (letter) by Darren Naish (7) 37
A Bibliography of Cryptozoological and Zoonythological Books by Dr Karl P.N.Shuker and Stephen Shipp (6) 26-29; (7) 30-35
Big Cats in the Garden of England by Neil Arnold (6) 13-14
Bird Brained (letter) by Eric Sorensen (7) 36
The Black Dogs of Dartmoor by Stephen Shipp (3) 12-14
Boars and Pumas by Roy Kerridge (5) 13-17
A Caledonian Collection (letter) by Tom Anderson (6) 31
The Case for Owlperson by Tony 'Doc' Shiels (6) 19-20
The Case of the Hairy Hands by Stephen Shipp (5) 26-27
A Caseful of Cougars by Jonathan Downes (7) 13-16
On Collecting a Cryptid by Grover S Krantz (7) 16-18
Creature from Clear lake - Catfish or Primitive Whale by Jan Williams (1) 6-7
Crocodile Tears by Jonathan Downes (4) 17-20
Crocodile Tears 2 by Jonathan Downes (5) 22
A Crypto Carol by Eric Sorenson (poem) (6) 30
Cryptocetology - introducing a new branch of Cryptozoology (Part 1) by Darren Naish (7) 19-27
The Doctor and the Owlman by Jonathan Downes (6) 15-16
Easy as ABC (letter) by Eric Sorensen (5) 31

Evidence for the hitherto unsuspected survival of two rare mustelids in the South West of England together with a re-appraisal of their taxonomic status by Jonathan Downes (1) 10-18
Expedition Report-In search of the Nepalese Humped Elephants (6) 11-12
Feathered Folklore - by Alison and Jonathan Downes (1) 9
Funky Frogs by Andy Stephens (5) 34
Giant 'Rabbits' in Devonshire by Jonathan Downes (1) 24-5
Golden Frogs of Bovey Tracey by Jonathan Downes (1) 21-2
Green Lizards in Devon and Dorset by Jonathan Downes (2) 22-23
A Hard Day's Night Parrot by Alison Downes (2) 21
If you go down to the woods today.... by Jan williams (2) 17-18
Is this 'Animal' Behaviour? by 'Paterfamilias (2) 19
King Kellas? by Tom Anderson (6) 29
The Last Word on the Beast of Bala (letter) by Alan Pringle (6) 30
Loch'ed in Combat/once again (two letters) by Mike Grayson and Nick Morgan
The Monster Mash (three letters on Loch Ness from G.M Stocker, R.A.Carter and Stephen Nice
The Migo Movie-a further muddying of muddy waters by Dr Karl P.N.Shuker (5) 22-25
Murder She Wrote by Alison Downes (1) 25
My sighting of Owlman by 'Gavin' (6) 16-16
The Mysterious Hominoids of Africa in the Light of Modern Research by François de Sarre (6) 23-25
Mystery Cats - where do we go from here by Jan Williams (3) 20-25
Near Lizard but not near enough by Stuart Leadbetter (2) 13-16
Near Lizard but not near enough - an addendum by Stuart Leadbetter (5) 28-29
Nepalese Humped Elephants (letter) by C.H.Keeling (7) 37
Ness than Zero by Petrovic (2) 23
Nigerian Scorpion Mystery (letter) by 'A retired Colonial Officer' (3) 26
A Norfolk Snarleygow by Jan Williams (3) 7-8
Not a lot of people know this...but Rabbits do (letter) by Roger Hutchings (3) 27
Obituaries of Gerald Durrell and Jane Bradley by Jonathan Downes (5) 35-36
On the bonny bonny banks (letter) by M.Playfair
An Opera worth Dying for (letter) by Lars Thomas
Oroborous is alive and well and living in Eastbourne by Sally Parsons and Jonathan Downes (3) 17
Out of Place and out on a limb by Alison Downes (1) 27
Packed Sabres by Eric Sorensen (6) 21-22
The Partridge Families Greatest Hits by Alison Downes (1) 27
A Question of Rhinoceri by Jonathan Downes (1) 28-9
The Saga of the St. Helena Sirenians by Dr karl P.N.Shuker (4) 12-16
Seabird Disasters by Alison Downes (1) 25
The search for Artrellia - the Papuan giant lizard by John Blashford-Snell (3) 15-18
Strange Snakes in Norfolk by Jan Williams (1) 24
Strangeness in Scotland by Mark Fraser (7) 28-30)
Three Million Cheers (letter) by Paul Nathan (4) 21-2
The Thylacine - the liveliest extinct animal around by Alan Pringle (4) 9-11
To Wit to Woo by Janet Bord (letter) (7) 36
Turning Japanese by Alison Downes (1) 27

Witness Reliability in Mystery Cat Reports - a cautionary tale by Jan Kingshott (5) 17-21
The Woodwose or Wildman in Britain by Ben Chapman (2) 9-12

****

Editor's Note:
This index is as accurate as we could make it. It must be stressed, however that it is not exhaustive, especially as regards part two and the editorial team are sure that much more exhaustive cross referencing as well as the inclusion of latin names and all places/names of eyewitnesses could have been included.
They were not, for reasons of space. The editorial team welcome comments and suggestions for the next index which will be included in the 1997 Yearbook.

1995 - A year in the life of the Centre for Fortean Zoology.

1995 was the year in which the Centre for Fortean Zoology finally stopped being merely a figment of my over-active imagination and finally took its place in Fortean and Zoological society.

1995 was the year that our journal, 'Animals & Men', started to sell enough for it to stop being merely a hobby and to justify its existence.

1995 was the year when Scott Walker, the avant garde torch singer, finally released a new record called 'Tilt'. It contained the lines: *"The good news you cannot refuse/the bad news is there is no news"*.

That couplet defined much of 1995 at The Centre for Fortean Zoology.

In January, the new Exeter radio station, Gemini Radio started to broadcast. One of the most important 'voices' of this new station is Australian DJ Steve Browning, a delightfully eccentric and massively entertaining character. Within a week or so of the station being in business I was a guest on his show talking about 'The Beast of Exmoor' and various hairy man-beasts. Over the next few months I clocked up about a dozen appearances on the Steve Browning show, and when the storm which followed the MAFF report about 'The Beast of Bodmin' broke in July/August my appearances on the show were a tangental part of the events that followed. (See A&M7).

In February the editorial team suffered a tragedy. Cartoonist Jane Bradley, a valued member of the CFZ team since the beginning and a good and dear friend of us all, was killed in a motor accident on the M5. this threw us badly, but we regrouped and released our second magazine of the year in April in time for the Fortean Times UNCONVENTION 1995.

The Unconvention was great fun and we made a number of valued contacts including veteran forteans Jerry Clark and Loren Coleman. We also made enough money to justify the first size increase of the year...from 32 to 40 pages. In October we were to carry out a second size increase to 44 pages.

As a result of our presence at the Unconvention I made a number of TV appearances to publicise our activities, and we managed to make the first documentary clips about such obscure beasties as 'the Golden Frogs of Bovey Tracey' and even 'The Owlman of Mawnan'. We also took TV cameras inside our favourite museum; 'Potter's Museum of Curiosity' on Bodmin Moor.

Much of the summer was, as the article in A&M7 reveals in greater depth, spent unravelling the truth, (or at least, some semblance of the truth), behind the discovery of a leopard skull on Bodmin Moor. We also saw 'Doc' again and he, (successfully as it turned out), predicted that both the Owlman and Morgawr would make return appearances during the summer and autumn.

In early September we put in an appearance at the Zoologica exhibition in Sussex, where, again we made a number of valuable contacts, and very bravely avoided the temptation of spending enormous sums of money that we could ill afford on some of the wonderful creatures which were on sale there. Those who know us and our propensity for purchasing strange beasts will be amazed to hear that not an animal did we buy!

Earlier in the year we had distributed a questionnaire asking people for their opinions of the CFZ and the service that we provide. It seems, from the results that, although most of the readers of 'Animals & Men' feel that its coverage and attitude is about right, a significant minority (about 22%) felt that we were not academic enough in our approach. This lead, directly to the publication of this book, the first of what we hope will be an annual event.

We have several other book projects in the pipeline, some of which will be published by outside publishers and some of which will be self published. Titles on the smaller mystery carnivores of the South-West and the Owlman of Mawnan and other feathered humanoid entities are presently in preparation.

During the year several of our new friends and colleagues came to visit. These included Sally Parsons, Richard Muirhead, Darren Naish and Daniel Bamping. It was very nice to see you all. I hope that our long term plans for establishing a proper visitors centre will finally reach fruition.

As we approach the end of the year, (I am writing this in October), the future looks exciting. Next year should be one for us all to look forward to.

Jonathan Downes.

THE CENTRE FOR FORTEAN ZOOLOGY

So, what is the Centre for Fortean Zoology?

We are a non profit-making organisation founded in 1992 with the aim of being a clearing house for information, and coordinating research into mystery animals around the world. We also study out of place animals, rare and aberrant animal behaviour, and Zooform Phenomena; – little-understood "things" that appear to be animals, but which are in fact nothing of the sort, and not even alive (at least in the way we understand the term).

Why should I join the Centre for Fortean Zoology?

Not only are we the biggest organisation of our type in the world but - or so we like to think - we are the best. We are certainly the only truly global Cryptozoological research organisation, and we carry out our investigations using a strictly scientific set of guidelines. We are expanding all the time and looking to recruit new members to help us in our research into mysterious animals and strange creatures across the globe. Why should you join us? Because, if you are genuinely interested in trying to solve the last great mysteries of Mother Nature, there is nobody better than us with whom to do it.

What do I get if I join the Centre for Fortean Zoology?

For £12 a year, you get a four-issue subscription to our journal *Animals & Men*. Each issue contains 60 pages packed with news, articles, letters, research papers, field reports, and even a gossip column! The magazine is A5 in format with a full colour cover. You also have access to one of the world's largest collections of resource material dealing with cryptozoology and allied disciplines, and people from the CFZ membership regularly take part in fieldwork and expeditions around the world.

How is the Centre for Fortean Zoology organized?

The CFZ is managed by a three-man board of trustees, with a non-profit making trust registered with HM Government Stamp Office. The board of trustees is supported by a Permanent Directorate of full and part-time staff, and advised by a Consultancy Board of specialists - many of whom who are world-renowned experts in their particular field. We have regional representatives across the UK, the USA, and many other parts of the world, and are affiliated with other organisations whose aims and protocols mirror our own.

I am new to the subject, and although I am interested I have little practical knowledge. I don't want to feel out of my depth. What should I do?

Don't worry. We were *all* beginners once. You'll find that the people at the CFZ are friendly and approachable. We have a thriving forum on the website which is the hub of an ever-growing electronic community. You will soon find your feet. Many members of the CFZ Permanent Directorate started off as ordinary members, and now work full time chasing monsters around the world.

I have an idea for a project which isn't on your website. What do I do?

Write to us, e-mail us, or telephone us. The list of future projects on the website is not exhaustive. If you have a good idea for an investigation, please tell us. We may well be able to help.

How do I go on an expedition?

We are always looking for volunteers to join us. If you see a project that interests you, do not hesitate to get in touch with us. Under certain circumstances we can help provide funding for your trip. If you look on the future projects section of the website, you can see some of the projects that we have pencilled in for the next few years.

In 2003 and 2004 we sent three-man expeditions to Sumatra looking for Orang-Pendek - a semi-legendary bipedal ape. The same three went to Mongolia in 2005. All three members started off merely subscribers to the CFZ magazine.

Next time it could be you!

Project Kerinci, Sumatra - 2003
In search of the bipedal ape Orang Pendek

How is the Centre for Fortean Zoology funded?

We have no magic sources of income. All our funds come from donations, membership fees, works that we do for TV, radio or magazines, and sales of our publications and merchandise. We are always looking for corporate sponsorship, and other sources of revenue. If you have any ideas for fund-raising please let us know. However, unlike other cryptozoological organisations in the past, we do not live in an intellectual ivory tower. We are not afraid to get our hands dirty, and furthermore we are not one of those organisations where the membership have to raise money so that a privileged few can go on expensive foreign trips. Our research teams both in the UK and abroad, consist of a mixture of experienced and inexperienced personnel. We are truly a community, and work on the premise that the benefits of CFZ membership are open to all.

What do you do with the data you gather from your investigations and expeditions?

Reports of our investigations are published on our website as soon as they are available. Preliminary reports are posted within days of the project finishing.

Each year we publish a 200 page yearbook containing research papers and expedition reports too long to be printed in the journal. We freely circulate our information to anybody who asks for it.

Is the CFZ community purely an electronic one?

No. Each year since 2000 we have held our annual convention - the *Weird Weekend* - in Exeter. It is three days of lectures, workshops, and excursions. But most importantly it is a chance for members of the CFZ to meet each other, and to talk with the members of the permanent directorate in a relaxed and informal setting and preferably with a pint of beer in one hand. Starting this year-18-20 August 2006 - the *Weird Weekend* will be bigger and better and held in the idyllic rural location of Woolsery in North Devon.

We are hoping to start up some regional groups in both the UK and the US which will have regular meetings, work together on research projects, and maybe have a mini convention of their own.

Since relocating to North Devon in 2005 we have become ever more closely involved with other community organisations, and we hope that this trend will continue. We also work closely with Police Forces across the UK as consultants for animal mutilation cases, and during 2006 we intend to forge closer links with the coastguard and other community services. We want to work closely with those who regularly travel into the Bristol Channel, so that if the recent trend of exotic animal visitors to our coastal waters continues, we can be out there as soon as possible.

We are building a Visitor's Centre in rural North Devon. This will not be open to the general public, but will provide a museum, a library and an educational resource for our members (currently over 400) across the globe. We are also planning a youth organisation which will involve children and young people in our activities.

Apart from having been the only Fortean Zoological organisation in the world to have consistently published material on all aspects of the subject for over a decade, we have achieved the following concrete results:

- Disproved the myth relating to the headless so-called sea-serpent carcass of Durgan beach in Cornwall 1975
- Disproved the story of the 1988 puma skull of Lustleigh Cleave
- Carried out the only in-depth research ever into mythos of the Cornish Owlma
- Made the first records of a tropical species of lamprey
- Made the first records of a luminous cave gnat larva in Thailand.
- Discovered a possible new species of British mammal - The Beech Marten.
- In 1994-6 carried out the first archival fortean zoological survey of Hong Kong.
- In the year 2000, CFZ theories where confirmed when an entirely new species of lizard was found resident in Britain.
- Identified the monster of Martin Mere in Lancashire as a giant wels catfish
- Expanded the known range of Armitage's skink in the Gambia by 80%
- Obtained photographic evidence of the remains of Europe's largest known pike
- Carried out the first ever in-depth study of the *ninki-nanka*
- Carried out the first attempt to breed Puerto Rican cave snails in captivity
- Were the first European explorers to visit the `lost valley` in Sumatra

EXPEDITIONS & INVESTIGATIOINS TO DATE INCLUDE

- 1998 Puerto Rico, Florida, Mexico *(Chupacabras)*
- 1999 Nevada *(Bigfoot)*
- 2000 Thailand *(Giant snakes called nagas)*
- 2002 Martin Mere *(Giant catfish)*
- 2002 Cleveland *(Wallaby mutilation)*
- 2003 Bolam Lake *(BHM Reports)*
- 2003 Sumatra *(Orang Pendek)*
- 2003 Texas *(Bigfoot; giant snapping turtles)*
- 2004 Sumatra *(Orang Pendek; cigau, a sabre-toothed cat)*
- 2004 Illinois *(Black panthers; cicada swarm)*
- 2004 Texas *(Mystery blue dog)*
- 2004 Puerto Rico *(Chupacabras; carnivorous cave snails)*
- 2005 Belize *(Affiliate expedition for hairy dwarfs)*
- 2005 Mongolia *(Allghoi Khorkhoi aka Mongolian death worm)*
- 2006 Gambia *(Gambo - Gambian sea monster , Ninki Nanka and Armitage s skink*
- 2006 Llangorse Lake *(Giant pike, giant eels)*
- 2006 Windermere *(Giant eels)*
- 2007 Coniston Water *(Giant eels)*
- 2007 Guyana *(Giant anaconda, didi, water tiger)*

To apply for a <u>FREE</u> information pack about the organisation and details of how to join, plus information on current and future projects, expeditions and events.

Send a stamped and addressed envelope to:

**THE CENTRE FOR FORTEAN ZOOLOGY
MYRTLE COTTAGE, WOOLSERY,
BIDEFORD, NORTH DEVON
EX39 5QR.**

or alternatively visit our website at:
www.cfz.org.uk

Other books available from
CFZ PRESS

CFZ PRESS

THE OWLMAN AND OTHERS - 30th Anniversary Edition
Jonathan Downes - ISBN 978-1-905723-02-7

£14.99

EASTER 1976 - Two young girls playing in the churchyard of Mawnan Old Church in southern Cornwall were frightened by what they described as a "nasty bird-man". A series of sightings that has continued to the present day. These grotesque and frightening episodes have fascinated researchers for three decades now, and one man has spent years collecting all the available evidence into a book. To mark the 30th anniversary of these sightings, Jonathan Downes has published a special edition of his book.

DRAGONS - More than a myth?
Richard Freeman - ISBN 0-9512872-9-X

£14.99

First scientific look at dragons since 1884. It looks at dragon legends worldwide, and examines modern sightings of dragon-like creatures, as well as some of the more esoteric theories surrounding dragonkind.

Dragons are discussed from a folkloric, historical and cryptozoological perspective, and Richard Freeman concludes that: "When your parents told you that dragons don't exist - they lied!"

MONSTER HUNTER
Jonathan Downes - ISBN 0-9512872-7-3

£14.99

Jonathan Downes' long-awaited autobiography, *Monster Hunter*...

Written with refreshing candour, it is the extraordinary story of an extraordinary life, in which the author crosses paths with wizards, rock stars, terrorists, and a bewildering array of mythical and not so mythical monsters, and still just about manages to emerge with his sanity intact.......

MONSTER OF THE MERE
Jonathan Downes - ISBN 0-9512872-2-2

£12.50

It all starts on Valentine's Day 2002 when a Lancashire newspaper announces that "Something" has been attacking swans at a nature reserve in Lancashire. Eyewitnesses have reported that a giant unknown creature has been dragging fully grown swans beneath the water at Martin Mere. An intrepid team from the Exeter based Centre for Fortean Zoology, led by the author, make two trips – each of a week – to the lake and its surrounding marshlands. During their investigations they uncover a thrilling and complex web of historical fact and fancy, quasi Fortean occurrences, strange animals and even human sacrifice.

**CFZ PRESS, MYRTLE COTTAGE,
WOOLFARDISWORTHY BIDEFORD,
NORTH DEVON, EX39 5QR
w w w . c f z . o r g . u k**

Other books available from
CFZ PRESS

ONLY FOOLS AND GOATSUCKERS
Jonathan Downes - ISBN 0-9512872-3-0

£12.50

In January and February 1998 Jonathan Downes and Graham Inglis of the Centre for Fortean Zoology spent three and a half weeks in Puerto Rico, Mexico and Florida, accompanied by a film crew from UK Channel 4 TV. Their aim was to make a documentary about the terrifying chupacabra - a vampiric creature that exists somewhere in the grey area between folklore and reality. This remarkable book tells the gripping, sometimes scary, and often hilariously funny story of how the boys from the CFZ did their best to subvert the medium of contemporary TV documentary making and actually do their job.

WHILE THE CAT'S AWAY
Chris Moiser - ISBN: 0-9512872-1-4

£7.99

Over the past thirty years or so there have been numerous sightings of large exotic cats, including black leopards, pumas and lynx, in the South West of England. Former Rhodesian soldier Sam McCall moved to North Devon and became a farmer and pub owner when Rhodesia became Zimbabwe in 1980. Over the years despite many of his pub regulars having seen the "Beast of Exmoor" Sam wasn't at all sure that it existed. Then a series of happenings made him change his mind. Chris Moiser—a zoologist—is well known for his research into the mystery cats of the westcountry. This is his first novel.

CFZ EXPEDITION REPORT 2006 - GAMBIA
ISBN 1905723032

£12.50

In July 2006, The J.T.Downes memorial Gambia Expedition - a six-person team - Chris Moiser, Richard Freeman, Chris Clarke, Oll Lewis, Lisa Dowley and Suzi Marsh went to the Gambia, West Africa. They went in search of a dragon-like creature, known to the natives as `Ninki Nanka`, which has terrorized the tiny African state for generations, and has reportedly killed people as recently as the 1990s. They also went to dig up part of a beach where an amateur naturalist claims to have buried the carcass of a mysterious fifteen foot sea monster named 'Gambo', and they sought to find the Armitage's Skink (*Chalcides armitagei*) - a tiny lizard first described in 1922 and only rediscovered in 1989. Here, for the first time, is their story.... With an forward by Dr. Karl Shuker and introduction by Jonathan Downes.

BIG CATS IN BRITAIN YEARBOOK 2006
Edited by Mark Fraser - ISBN 978-1905723-01-0

£10.00

Big cats are said to roam the British Isles and Ireland even now as you are sitting and reading this. People from all walks of life encounter these mysterious felines on a daily basis in every nook and cranny of these two countries. Most are jet-black, some are white, some are brown, in fact big cats of every description and colour are seen by some unsuspecting person while on his or her daily business. 'Big Cats in Britain' are the largest and most active group in the British Isles and Ireland This is their first book. It contains a run-down of every known big cat sighting in the UK during 2005, together with essays by various luminaries of the British big cat research community which place the phenomenon into scientific, cultural, and historical perspective.

CFZ PRESS, MYRTLE COTTAGE,
WOOLSERY, BIDEFORD,
NORTH DEVON, EX39 5QR
w w w . c f z . o r g . u k

Other books available from
CFZ PRESS

THE SMALLER MYSTERY CARNIVORES OF THE WESTCOUNTRY
Jonathan Downes - ISBN 978-1-905723-05-8

£7.99

Although much has been written in recent years about the mystery big cats which have been reported stalking Westcountry moorlands, little has been written on the subject of the smaller British mystery carnivores. This unique book redresses the balance and examines the current status in the Westcountry of three species thought to be extinct: the Wildcat, the Pine Marten and the Polecat, finding that the truth is far more exciting than the currently held scientific dogma. This book also uncovers evidence suggesting that even more exotic species of small mammal may lurk hitherto unsuspected in the countryside of Devon, Cornwall, Somerset and Dorset.

THE BLACKDOWN MYSTERY
Jonathan Downes - ISBN 978-1-905723-00-3

£7.99

Intrepid members of the CFZ are up to the challenge, and manage to entangle themselves thoroughly in the bizarre trappings of this case. This is the soft underbelly of ufology, rife with unsavoury characters, plenty of drugs and booze." That sums it up quite well, we think. A new edition of the classic 1999 book by legendary fortean author Jonathan Downes. In this remarkable book, Jon weaves a complex tale of conspiracy, anti-conspiracy, quasi-conspiracy and downright lies surrounding an air-crash and alleged UFO incident in Somerset during 1996. However the story is much stranger than that. This excellent and amusing book lifts the lid off much of contemporary forteana and explains far more than it initially promises.

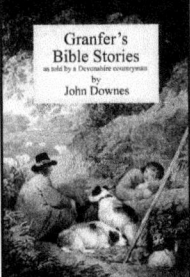

GRANFER'S BIBLE STORIES
John Downes - ISBN 0-9512872-8-1

£7.99

Bible stories in the Devonshire vernacular, each story being told by an old Devon Grandfather - 'Granfer'. These stories are now collected together in a remarkable book presenting selected parts of the Bible as one more-or-less continuous tale in short 'bite sized' stories intended for dipping into or even for bed-time reading. `Granfer` treats the biblical characters as if they were simple country folk living in the next village. Many of the stories are treated with a degree of bucolic humour and kindly irreverence, which not only gives the reader an opportunity to re-evaluate familiar tales in a new light, but do so in both an entertaining and a spiritually uplifting manner.

FRAGRANT HARBOURS DISTANT RIVERS
John Downes - ISBN 0-9512872-5-7

£12.50

Many excellent books have been written about Africa during the second half of the 19th Century, but this one is unique in that it presents the stories of a dozen different people, whose interlinked lives and achievements have as many nuances as any contemporary soap opera. It explains how the events in China and Hong Kong which surrounded the Opium Wars, intimately effected the events in Africa which take up the majority of this book. The author served in the Colonial Service in Nigeria and Hong Kong, during which he found himself following in the footsteps of one of the main characters in this book; Frederick Lugard – the architect of modern Nigeria.

CFZ PRESS, MYRTLE COTTAGE, WOOLFARDISWORTHY BIDEFORD, NORTH DEVON, EX39 5QR
w w w . c f z . o r g . u k

Other books available from
CFZ PRESS

ANIMALS & MEN - Issues 1 - 5 - In the Beginning
Edited by Jonathan Downes - ISBN 0-9512872-6-5

£12.50

At the beginning of the 21st Century monsters still roam the remote, and sometimes not so remote, corners of our planet. It is our job to search for them. The Centre for Fortean Zoology [CFZ] is the only professional, scientific and full-time organisation in the world dedicated to cryptozoology - the study of unknown animals. Since 1992 the CFZ has carried out an unparalleled programme of research and investigation all over the world. We have carried out expeditions to Sumatra (2003 and 2004), Mongolia (2005), Puerto Rico (1998 and 2004), Mexico (1998), Thailand (2000), Florida (1998), Nevada (1999 and 2003), Texas (2003 and 2004), and Illinois (2004). An introductory essay by Jonathan Downes, notes putting each issue into a historical perspective, and a history of the CFZ.

ANIMALS & MEN - Issues 6 - 10 - The Number of the Beast
Edited by Jonathan Downes - ISBN 978-1-905723-06-5

£12.50

At the beginning of the 21st Century monsters still roam the remote, and sometimes not so remote, corners of our planet. It is our job to search for them. The Centre for Fortean Zoology [CFZ] is the only professional, scientific and full-time organisation in the world dedicated to cryptozoology - the study of unknown animals. Since 1992 the CFZ has carried out an unparalleled programme of research and investigation all over the world. We have carried out expeditions to Sumatra (2003 and 2004), Mongolia (2005), Puerto Rico (1998 and 2004), Mexico (1998), Thailand (2000), Florida (1998), Nevada (1999 and 2003), Texas (2003 and 2004), and Illinois (2004). Preface by Mark North and an introductory essay by Jonathan Downes, notes putting each issue into a historical perspective, and a history of the CFZ.

BIG BIRD! Modern Sightings of Flying Monsters

Ken Gerhard - ISBN 978-1-905723-08-9

£7.99

From all over the dusty U.S./Mexican border come hair-raising stories of modern day encounters with winged monsters of immense size and terrifying appearance. Further field sightings of similar creatures are recorded from all around the globe. What lies behind these weird tales? Ken Gerhard is a native Texan, he lives in the homeland of the monster some call 'Big Bird'. Ken's scholarly work is the first of its kind. On the track of the monster, Ken uncovers cases of animal mutilations, attacks on humans and mounting evidence of a stunning zoological discovery ignored by mainstream science. Keep watching the skies!

STRENGTH THROUGH KOI
They saved Hitler's Koi and other stories

Jonathan Downes - ISBN 978-1-905723-04-1

£7.99

Strength through Koi is a book of short stories - some of them true, some of them less so - by noted cryptozoologist and raconteur Jonathan Downes. The stories are all about koi carp, and their interaction with bigfoot, UFOs, and Nazis. Even the late George Harrison makes an appearance. Very funny in parts, this book is highly recommended for anyone with even a passing interest in aquaculture, but should be taken definitely *cum grano salis*.

CFZ PRESS, MYRTLE COTTAGE, WOOLSERY, BIDEFORD, NORTH DEVON, EX39 5QR

Other books available from
CFZ PRESS

BIG CATS IN BRITAIN YEARBOOK 2007
Edited by Mark Fraser - ISBN 978-1-905723-09-6

£12.50

People from all walks of life encounter mysterious felids on a daily basis, in every nook and cranny of the UK. Most are jet-black, some are white, some are brown; big cats of every description and colour are seen by some unsuspecting person while on his or her daily business. 'Big Cats in Britain' are the largest and most active research group in the British Isles and Ireland. This book contains a run-down of every known big cat sighting in the UK during 2006, together with essays by various luminaries of the British big cat research community.

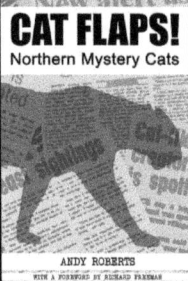

CAT FLAPS! Northern Mystery Cats
Andy Roberts - ISBN 978-1-905723-11-9

£6.99

Of all Britain`s mystery beasts, the alien big cats are the most renowned. In recent years the notoriety of these uncatchable, out-of-place predators have eclipsed even the Loch Ness Monster. They slink from the shadows to terrorise a community, and then, as often as not, vanish like ghosts. But now film, photographs, livestock kills, and paw prints show that we can no longer deny the existence of these once-legendary beasts. Here then is a case-study, a true lost classic of Fortean research by one of the country's most respected researchers.

CENTRE FOR FORTEAN ZOOLOGY 2007 YEARBOOK
Edited by Jonathan Downes and Richard Freeman
ISBN 978-1-905723-14-0

£12.50

The Centre For Fortean Zoology Yearbook is a collection of papers and essays too long and detailed for publication in the CFZ Journal *Animals & Men*. With contributions from both well-known researchers, and relative newcomers to the field, the Yearbook provides a forum where new theories can be expounded, and work on little-known cryptids discussed.

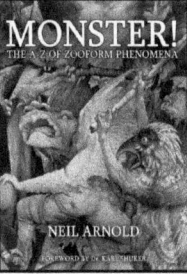

MONSTER! THE A-Z OF ZOOFORM PHENOMENA
Neil Arnold - ISBN 978-1-905723-10-2

£14.99

Zooform Phenomena are the most elusive, and least understood, mystery `animals`. Indeed, they are not animals at all, and are not even animate in the accepted terms of the word. Author and researcher Neil Arnold is to be commended for a groundbreaking piece of work, and has provided the world's first alphabetical listing of zooforms from around the world.

**CFZ PRESS, MYRTLE COTTAGE,
WOOLFARDISWORTHY BIDEFORD,
NORTH DEVON, EX39 5QR
www.cfz.org.uk**

Other books available from
CFZ PRESS

BIG CATS LOOSE IN BRITAIN
Marcus Matthews - ISBN 978-1-905723-12-6

£14.99

Big Cats: Loose in Britain, looks at the body of anecdotal evidence for such creatures: sightings, livestock kills, paw-prints and photographs, and seeks to determine underlying commonalities and threads of evidence. These two strands are repeatedly woven together into a highly readable, yet scientifically compelling, overview of the big cat phenomenon in Britain.

DARK DORSET
TALES OF MYSTERY, WONDER AND TERROR
Robert. J. Newland and Mark. J. North
ISBN 978-1-905723-15-6

£12.50

This extensively illustrated compendium has over 400 tales and references, making this book by far one of the best in its field. Dark Dorset has been thoroughly researched, and includes many new entries and up to date information never before published. The title of the book speaks for itself, and is indeed not for the faint hearted or those easily shocked.

MAN-MONKEY - IN SEARCH OF THE BRITISH BIGFOOT
Nick Redfern - ISBN 978-1-905723-16-4

£9.99

In her 1883 book, *Shropshire Folklore*, Charlotte S. Burne wrote: *'Just before he reached the canal bridge, a strange black creature with great white eyes sprang out of the plantation by the roadside and alighted on his horse's back'*. The creature duly became known as the `Man-Monkey`.

Between 1986 and early 2001, Nick Redfern delved deeply into the mystery of the strange creature of that dark stretch of canal. Now, published for the very first time, are Nick's original interview notes, his files and discoveries; as well as his theories pertaining to what lies at the heart of this diabolical legend.

EXTRAORDINARY ANIMALS REVISITED
Dr Karl Shuker - ISBN 978-1905723171

£14.99

This delightful book is the long-awaited, greatly-expanded new edition of one of Dr Karl Shuker's much-loved early volumes, *Extraordinary Animals Worldwide*. It is a fascinating celebration of what used to be called romantic natural history, examining a dazzling diversity of animal anomalies, creatures of cryptozoology, and all manner of other thought-provoking zoological revelations and continuing controversies down through the ages of wildlife discovery.

**CFZ PRESS, MYRTLE COTTAGE,
WOOLFARDISWORTHY BIDEFORD,
NORTH DEVON, EX39 5QR
www.cfz.org.uk**

Other books available from
CFZ PRESS

DARK DORSET CALENDAR CUSTOMS
Robert J Newland - ISBN 978-1-905723-18-8

£12.50

Much of the intrinsic charm of Dorset folklore is owed to the importance of folk customs. Today only a small amount of these curious and occasionally eccentric customs have survived, while those that still continue have, for many of us, lost their original significance. Why do we eat pancakes on Shrove Tuesday? Why do children dance around the maypole on May Day? Why do we carve pumpkin lanterns at Hallowe'en? All the answers are here! Robert has made an in-depth study of the Dorset country calendar identifying the major feast-days, holidays and celebrations when traditionally such folk customs are practiced.

CENTRE FOR FORTEAN ZOOLOGY 2004 YEARBOOK
Edited by Jonathan Downes and Richard Freeman
ISBN 978-1-905723-14-0

£12.50

The Centre For Fortean Zoology Yearbook is a collection of papers and essays too long and detailed for publication in the CFZ Journal *Animals & Men*. With contributions from both well-known researchers, and relative newcomers to the field, the Yearbook provides a forum where new theories can be expounded, and work on little-known cryptids discussed.

CENTRE FOR FORTEAN ZOOLOGY 2008 YEARBOOK
Edited by Jonathan Downes and Corinna Downes
ISBN 978 -1-905723-19-5

£12.50

The Centre For Fortean Zoology Yearbook is a collection of papers and essays too long and detailed for publication in the CFZ Journal *Animals & Men*. With contributions from both well-known researchers, and relative newcomers to the field, the Yearbook provides a forum where new theories can be expounded, and work on little-known cryptids discussed.

ETHNA'S JOURNAL
Corinna Newton Downes
ISBN 978 -1-905723-21-8

£9.99

Ethna's Journal tells the story of a few months in an alternate Dark Ages, seen through the eyes of Ethna, daughter of Lord Edric. She is an unsophisticated girl from the fortress town of Cragnuth, somewhere in the north of England, who reluctantly gets embroiled in a web of treachery, sorcery and bloody war...

**CFZ PRESS, MYRTLE COTTAGE,
WOOLFARDISWORTHY BIDEFORD,
NORTH DEVON, EX39 5QR
w w w . c f z . o r g . u k**

Other books available from
CFZ PRESS

CFZ PRESS

ANIMALS & MEN - Issues 11 - 15 - The Call of the Wild
Jonathan Downes (Ed) - ISBN 978-1-905723-07-2

£12.50

Since 1994 we have been publishing the world's only dedicated cryptozoology magazine, *Animals & Men*. This volume contains fascimile reprints of issues 11 to 15 and includes articles covering out of place walruses, feathered dinosaurs, possible North American ground sloth survival, the theory of initial bipedalism, mystery whales, mitten crabs in Britain, Barbary lions, out of place animals in Germany, mystery pangolins, the barking beast of Bath, Yorkshire ABCs, Molly the singing oyster, singing mice, the dragons of Yorkshire, singing mice, the bigfoot murders, waspman, British beavers, the migo, Nessie, the weird warbling whatsit of the westcountry, the quagga project and much more...

IN THE WAKE OF BERNARD HEUVELMANS
Michael A Woodley - ISBN 978-1-905723-20-1

£9.99

Everyone is familiar with the nautical maps from the middle ages that were liberally festooned with images of exotic and monstrous animals, but the truth of the matter is that the *idea* of the sea monster is probably as old as humankind itself.

For two hundred years, scientists have been producing speculative classifications of sea serpents, attempting to place them within a zoological framework. This book looks at these successive classification models, and using a new formula produces a sea serpent classification for the 21st Century.

CFZ PRESS, MYRTLE COTTAGE,
WOOLFARDISWORTHY BIDEFORD,
NORTH DEVON, EX39 5QR
w w w . c f z . o r g . u k